THE GEOGRAPHICAL IMAGINATION IN AMERICA
1880–1950

☆

THE
GEOGRAPHICAL IMAGINATION
IN AMERICA, 1880–1950

☆

Susan Schulten

THE UNIVERSITY OF CHICAGO PRESS
CHICAGO AND LONDON

SUSAN SCHULTEN is assistant professor in the history department
at the University of Denver.

The University of Chicago Press, Chicago 60637
The University of Chicago Press, Ltd., London
© 2001 by The University of Chicago
All rights reserved. Published 2001
Printed in the United States of America
10 09 08 07 06 05 04 03 02 01   1 2 3 4 5

ISBN: 0-226-74055-2 (cloth)
ISBN: 0-226-74056-0 (paper)

An earlier version of chapter 2 appeared as "The Making of the Geographic:
Science, Culture, and Expansionism," *American Studies* 41, no. 1 (spring 2000):
5–29; © Mid-America American Studies Association, reprinted by permission.
A portion of chapter 7 appeared as "The Limits of Possibility: Rand McNally in
American Culture, 1898–1929," *Cartographic Perspectives* no. 35 (winter 2000):
7–26; reprinted by permission of the journal. A portion of chapter 8 appeared
as "Richard Edes Harrison and the Challenge to American Cartography," *Imago
Mundi: The International Journal for the History of Cartography* 50 (1998): 174–88;
reprinted by permission of the journal.

Library of Congress Cataloging-in-Publication Data

Schulten, Susan.
    The geographical imagination in America, 1880–1950 / Susan Schulten.
    p.   cm.
    Includes bibliographical references (p.) and index.
    ISBN 0-226-74055-2 (alk. paper)—ISBN 0-226-74056-0 (pbk. : alk. paper)
    1. Geography—History. 2. Geography—United States—History.
    3. Geography—Social aspects—United States. 4. Geography—Philosophy.
    5. United States—History. I. Title.
    G96 .S36   2001
    917—dc21
                                                            00-010159

CONTENTS

ILLUSTRATIONS

ACKNOWLEDGMENTS

M any individuals helped me complete this book. Archivists and librarians who deserve special mention include Ronald Grim and Jim Flatnes at the Library of Congress; Mary Anne McMillen at the National Geographic Society; Pat Morris and Robert Karrow at the Newberry Library; David Cobb and Joseph Garver at Harvard University's Pusey Library; and Steve Fisher at the University of Denver's Penrose Library. The formidable task of collecting and reproducing the illustrations was aided immensely by the staff at the Still Pictures Branch of the National Archives; Carmella Napoleone, Stephen Sylvester, and Robert Zinck at Harvard University's Imaging Services; Martin von Wyss at Harvard University's Map Collection; Maureen Dilg and Beth Rae Richardson at the National Geographic Society; Jil Dawicki, Tonya Yada, Diane Kotowski, and Justin Carricaburu at the University of Denver; John Powell at the Newberry Library; Bette Weneck at the Milbank Memorial Library of Teacher's College, Columbia University; Chelo and Rusty Pallas at Pallas Photo Imaging; and Elisabeth Filar at the University of Colorado's Map Library. I owe a debt to the Office of the Dean at the University of Denver: costs incurred by the illustrations in this book were offset by grants through the Walter Rosenberry Fund, and a short sabbatical awarded in fall 1998 gave me time to concentrate on writing. The university's Office of Sponsored Programs also contributed generously to the production of this book through its Faculty Research Fund.

Portions of this manuscript were read—and immensely improved—by Bruce Kuklick, Bin Ramke, Elliott Gorn, Ari Kelman, and the late John Livingston. For their interest in reading and shaping earlier formulations of my ideas I am grateful to Drew Faust, Walter McDougall, Michael Zuckerman, Catherine Murdock, Jon Leverenz, Gary Gerstle, and Charles Rosenberg. I have learned a great deal from historically minded geographers, especially Matthew Edney, David Woodward, Felix Driver, Jim Akerman, Neil Smith, David N. Livingstone, Tamar Rothenberg, and Roger Downs. More than a few historians have also helped me to contextualize geography, in particular Stephen Kern, Thomas Glick, Sumathi Ramaswamy,

Jeremy Black, and Alan Henrikson. The members of Jim Akerman's 1998 "Maps and Nations" seminar at the Newberry Library kindly responded to my ideas and helped push them in new directions. Robert Anderson and Mark Monmonier, as well as the anonymous readers at the University of Chicago Press, read the entire manuscript, and I am indebted to them for their patience and insight. The editorial assistance and guidance of Robert Devens, Doug Mitchell, Evan Young, and Leslie Keros was indispensable.

As my ideas began to take shape in Philadelphia, I was encouraged by conversations with Catherine and James Murdock, Max Page, Steve Conn, Tom Sugrue, Dana Barron, Rob Gregg, Frank Gavin, Nick Breyfogle, and Charlie Montgomery. I found a generous friend and roommate in Colleen McCauley, who greatly eased the solitary nature of research and writing. Over the years, my research visits to the Newberry Library in Chicago were enriched substantially by discussions with Elliott Gorn, Tim Gilfoyle, Jim Grossman, and Robin Bachin. My colleagues in the history department at the University of Denver have been especially supportive in the past few years, in particular Joyce Goodfriend, Michael Gibbs, Ingrid Tague, Carol Helstosky, and the late John Livingston. It is my good fortune to have made quick and lasting friendships in Denver, and for their enduring interest and involvement in this project I especially want to thank Liz Harris, Dan Feehan, David and Nell London, Susan Sterett, Michael Bennet, Susan Daggett, Ann Lininger, David White, and Howard Pincus. Vonnie and Warren Schulten have been indefatigable in their support; they along with Mark, David, and Amber Schulten have indulged my distracted mind and long silences with grace, warmth, and good humor. Robert Anderson has lived with this project in all its stages, and has given up a great deal to help me finish it. I am grateful for the quiet strength and unwavering love he has shown me in our years together.

1

Introduction

In 1923 Rand McNally's chief cartographer drew a new map of the world. John Paul Goode designed the "homolosine projection" in order to challenge the distortions perpetuated by the Mercator projection, a map that had long dominated American understandings of geography (figs. 1.1, 1.2). Goode's new map was welcomed at the annual meeting of professional geographers and eagerly hailed as a breakthrough in modern cartography. But while cartographers lauded the homolosine projection as technically superior to Mercator's, its general reception fell short of expectations. Many employees at Rand McNally were put off by its jagged, disorienting configuration. One expressed shock at "a world so grievously sundered."[1] Even company president Andrew McNally II worried that while the map might sell to schoolteachers—who were concerned with correcting the flaws of the Mercator projection—it was insufficiently "unified" in appearance to be accepted by the American public. Thus Rand McNally decided to market Goode's more "accurate" map to schools but declined to adopt it in the more popular, mass-market atlases. The company continued to rely on the Mercator projection because of its narrative power. Though most Americans had never heard of Gerhard Mercator, his sixteenth-century projection had in fact come to be taken *for* the world.

Mercator's map would not be widely challenged until World War II, when more Americans came into contact with maps than in any other period of American history. At three pivotal moments in the war—after the German invasion of Poland, the bombing of Pearl Harbor, and the assault on Normandy—Americans bought in a matter of hours what in peacetime would have been a year's supply of maps and atlases. This nationwide attention to maps brought the farthest reaches of the war—from

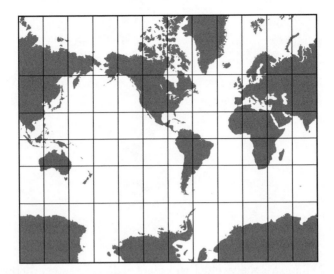

Figure 1.1   Mercator projection

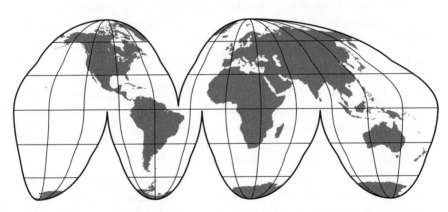

Figure 1.2   Goode's homolosine projection. Notice the fundamental difference in the "shape" of the world in this projection, designed in the interwar era, as compared to the long-accepted Mercator projection.

Guadalcanal to Anzio, Simferopol to Midway—into everyday conversation, and demonstrated the powerful relationship between war and geography. But even more compelling than the growing interest in distant lands is the effect the war had on the very *nature* of geographic knowledge and spatial representation. The countless maps Americans bought showed them a world that would have been utterly unfamiliar a decade earlier. Gone was Mercator's ordered plane, which had comfortably distanced Americans from Europe and Asia. The global nature of the war, together with the advent of aviation, completely reconfigured the look and shape of the world on a map. Americans now pored over maps that represented the world as a sphere, or that placed the North Pole at the center, projections and perspectives that would otherwise have been familiar only to cartographers. These new maps emphasized America's *proximity* to Europe and Asia over the North Pole and across the oceans, shaking the nation's well-developed sense of isolation. So powerful were these new views of the world that by the 1940s they had become symbols of internationalism and global interdependence (figs. 1.3, 1.4).

The "rise and fall" of the Mercator projection suggests the malleable nature of geographical knowledge. Taken more broadly, it presents a historical problem: how political, cultural, and social imperatives have shaped ideas about geography and space, and how these ideas have in turn influenced American history and culture. In essence, geography is a way of distinguishing "here" from "there," without which little sense can be made of human experience. Yet, as Martin Lewis and Kären Wigen have observed, the use of spatial structures to order the world can produce a kind of reflexive metageography, such as the identification of Europe and Asia as two separate continents, or the postcolonial framework of First, Second, and Third Worlds.[2] Buttressed by the authority of science, geographical knowledge casts off myths and half-truths for the concrete and coherent narration of place, and in so doing has the power to transform historical practices into transhistorical truths. From the isolationism of the 1880s to the internationalism of the 1940s, three overlapping traditions of geography— academic and school geography, mass-market cartography, and the work of the National Geographic Society—helped Americans imagine and comprehend a world that most would not experience firsthand. How these institutions and traditions were created, negotiated, entrenched, and occasionally challenged, and what interests they served, is the subject of this book. Part 1 examines the rise of these traditions of geographical knowledge in the late nineteenth and early twentieth centuries; part 2 asks how they struggled to make sense of an increasingly complicated world in which the role of the United States was rapidly expanding. This general chronological division is subdivided into a series of thematic chapters, each of which presents a problem of geographical understanding. Because this is a complex story that weaves together many disparate threads, there is at times overlap and some repetition across chapters. I hope readers will indulge these occasional departures and trust that they are included in the interest of a larger narrative coherence.

Figures 1.3, 1.4 Images such as these became symbols of the reimagined world and of American internationalism in the mid-twentieth century. From the Consolidated Vultee Aircraft Corporation.

Maps and the Language of American Power

The thought of geography evokes maps, the two-dimensional representations of the landscape that create order out of chaos. But while maps are arguments that mediate our understanding of the world, historians have paid relatively little attention to their role in American culture. Until recently, even many historians of cartography essentially regarded maps as scientific records of an expanding body of knowledge, driven by events such as the discovery of America, the exploration of Africa, or surveys of the American West. Many have asked how accurately maps reflect the terrain, so as they become more empirically verifiable they lose their mystery and interest to scholars and collectors alike. As a result, modern maps — produced after the peak of European exploration — tend to be treated as increasingly scientific reflections of the environment rather than as products of historical circumstance. Correspondingly, interest in maps produced during the twentieth century has largely focused on propaganda maps, which has engendered a rather murky distinction between objective and subjective cartography.[3] Yet contesting the accuracy of a map is a qualitatively different enterprise from dissecting its rhetorical structure. Because maps exist at the crossroads of science and culture, the issue is not simply one of precision, but "precision with respect to what?"[4] More recently, scholars have begun to challenge the assumption of maps as simply scientific and instead to ask broader questions about their function as arbiters of power. Brian Harley's concern with the map as a text has attracted the attention of scholars from a range of disciplines. The multivolume *History of Cartography* project reflects this critical turn within the history of cartography, and the growing interest in spatial concepts more generally. This critical, historical approach has yielded excellent studies of the cultural construction of cartography in Weimar and Nazi Germany, British India, and modern Thailand.[5]

In a similar spirit, this study explores the historical context of maps and mapping in American society. But what maps can tell us about American culture raises the more elusive question of how they shape and reflect popular ideas and attitudes. The power of maps to condition consciousness can easily be exaggerated when we consider that laments about American geographic ignorance were as common a century ago as they are today. On balance, however, these maps are worth studying for a number of reasons. First, the sheer number of Americans exposed to maps dramatically increased between 1880 and 1950, with explosions of interest during the Spanish-American War, World War I, and World War II. The laborious mapmaking methods used prior to the Civil War yielded a limited number of atlases that were priced well beyond the grasp of most Americans. By the 1870s, however, print technologies adopted by American mapmakers transformed cartography from an elite craft into a mass-production industry. The proliferation of maps through American culture in subsequent decades formed a normative — if passive — picture of the world, which is itself significant. As Brian Harley wrote, "the power of the mapmaker was not generally exercised over individuals but over the knowledge of the world made available to people in general."[6]

This "power of the mapmaker" to both reveal and conceal deserves special consideration in the American context. From the 1880s to the 1940s four of the largest map producers — Rand McNally, George F. Cram, C. S. Hammond, and the National Geographic Society — controlled a large share of the market, supplying maps and atlases not only to the public but also to the government, which had no substantial mapping agencies of its own until the eve of the Second World War. That each of these organizations was founded in the early era of mass production cartography is important. With only a few producers using relatively similar techniques, American maps began to develop a rather homogeneous style, one that became entrenched as the industry focused increasingly on profits rather than aesthetics, and tailored its products to suit the widest possible audience. Thus the industry designed maps that would appear familiar to consumers, and rarely challenged their assumptions about the look and shape of the world. Given the stark absence of diversity, we ought to consider how maps became more powerful as they replicated existing notions of how the landscape should appear. In an era of such little change, the commercially successful atlas *confirmed* widely held notions about the world. Put another way, the homogeneity of the map and the atlas — across the industry and over time — delimited the nation's sense of what was cartographically possible. Only when the cataclysms of a global air war forcibly disrupted conventional ideas about geography did the industry reconceptualize its understanding of the world on a map.

The importance of interpreting the map extends to the larger medium where most maps are found: the atlas. Until recently, cartographic historians had rarely treated atlases as deliberately conceived and composed texts, and there is a long tradition among collectors of dismembering atlases for the purpose of trading individual maps, a practice that treats the fact of the atlas as incidental. There is also — strangely — an inverse relationship between the proliferation of atlases in American culture and the level of critical attention paid to them, indicative of the low regard in which mass-market cartography is held by both scholars and collectors. Yet as the dissemination of atlases and maps steadily increased after the 1870s so too did their position within society, which makes scholarly attention correspondingly more, not less, important. The atlas contained not just maps but narrative descriptions, illustrations, graphs, and charts to bring cosmopolitanism and nationalism into America homes. Though historians cannot consistently trace how Americans read these texts, we can ask how they were produced and marketed and the rate at which they were consumed and absorbed into the culture. Analyzing how the atlas divided, organized, mapped, and narrated the world reveals how it purported to explain and contain the world for public consumption. As we will see, America's rise to internationalism shaped, and was facilitated by, popular cartography.

On balance, critical changes came to cartographic culture after the Spanish-American War and the Second World War, both periods that augmented national commitments abroad and revised American visions of the world. In each case the frenzy of war and the subsequent reorientation of foreign policy allowed geography and cartography to enter not just official discussions but public discourse as well, which eventually affected the design of atlases. Yet the history of American cartography is not simply, or even largely, a story of change. Perhaps even more important are the ideas and arguments that are powerful precisely because they appear to be natural, scientific, and unmediated representations of space that remain unchallenged over time. Through the study of company records, advertising, maps, and atlases, chapters 2, 8, and 9 investigate the power of cartography to create a world that implicitly endorsed the nation's aggregation of territory, resources, and capital.

NATIONAL GEOGRAPHIC AND THE CULTURE OF CURIOSITY

The history of the National Geographic Society, like that of maps, underscores the complex relationship between geographical knowledge and power. The Society was founded in 1888 to serve scientists working for the federal government through lectures and research published in the *National Geographic*. Initially the journal focused on the mundane work of federal surveys, geology, hydrography, and exploration, and conceived of geography primarily as a physical science related to domestic concerns. But the flurry of activity abroad at century's end gave the Society the rather unexpected opportunity to cover the "geographical" dimension of the Spanish-American War, and thereby bring the exotic reaches of the nation's new territories home to Americans. This goal reoriented the Society's idea of geography and encouraged it to use its scientific expertise to serve the state in new ways. Given their historically close relationship with federal bu-reaucracies, the editors had little difficulty transforming the *Geographic* from a staid record of government surveys into a vigorous advocate for inter-vention in the Pacific and the Caribbean. In this regard, geography grew up around national imperatives.

The modern form of the *Geographic* was born in the turmoil of war. In order to cover the war the magazine broadened its use of photography to in-clude human subjects, a tradition that has become one of the enduring pil-lars of its success. Filling its pages with pictures of Filipinos and Cubans—illustrations that eventually became even more central than the text—the magazine hit upon a formula that resonated deeply with its readers, one that fused entertainment with science. Soon thereafter the arrival of a new editor, drawn from outside government science, catalyzed the *Geographic's* turn toward its modern niche as a wildly successful medium of middlebrow culture. Gilbert Grosvenor's contribution stemmed in large part from his paradoxical ideal for the magazine. In 1915, with little sense of the implicit contradiction, Grosvenor declared that the *Geographic* would print only the most accurate information but at the same time would avoid anything neg-ative. Grosvenor believed that such an ideal would allow the *Geographic* to transcend the muck of politics and bias, and to move away from the crudely nationalistic posture that had characterized it at the turn of the century. Yet America's entry into World War I—combined with the *Geographic's* sensibility—turned the magazine away from Grosvenor's ideal in short or-der. Even before Congress committed the nation to war, the *Geographic* was vigorously defending the British and French against the "Huns" of Ger-many, a posture that brought the National Geographic Society censure and temporarily slowed membership growth. This experience renewed Grosve-nor's commitment to bringing readers only the bright side of life. And Grosvenor equated this buoyancy with accuracy, for what the magazine's readers wanted, he believed, was "mental relaxation without emotional stimulus."[7] What Grosvenor sought was a world without politics—a strat-egy that eventually led, absurdly, to the *Geographic's* enthusiastic portray-als of fascist regimes in the 1930s.

With a scope of "the world and all that is in it," the Society claimed more than one million members by 1926. Americans took their member-ship seriously, as evidenced by the vast correspondence sent to the organi-zation each year. As one woman wrote, "I don't know when I have been so proud of anything as the little certificate which I hold as a member of the Society. It is second to my college diploma."[8] These letters allow us to see how the Society negotiated the desires and concerns of its audience, and provide some clues to the contemporary "readings" of the *Geographic*. These letters also suggest that although Americans had diverse ideas about the goal and purpose of their membership, they considered it a shared intellectual enterprise and a mark of respectability. This sense of the magazine as a symbol of cosmopolitanism reveals how the *Geographic* was able to maintain its reputation as a source of wholesome fare while simultaneously printing photographs that—in any other context—would violate social mores. Few of the Society's members protested the magazine's trademark photographs of seminude subjects, because to take the *Geographic* was to identify oneself as curious but also sophisticated, someone who viewed photographs of "natives" not as a voyeur but as an amateur anthropologist.

Whether "discovering" the mountains of Machu Picchu or the natives of the Philippines, the *Geographic* was able to tap American curiosity about the farthest reaches of the earth. Catherine Lutz and Jane Collins have argued that, through photography, the *Geographic* gave Americans a thoroughly humanistic but fundamentally ahistorical reading of the world.[9] The *Geographic*, they observe, framed the world as "timeless, classless, and outside the boundaries of language and culture." Yet a more complete reading of the magazine requires a study of the Society and its early leaders, especially Alexander Graham Bell, W J McGee, and Gilbert Grosvenor. Chapters 3 and 7 do not constitute an exhaustive history of the Society from its founding through the 1920s, but they do ask how it came to be one of the most ubiquitous sources of information and images about the world in American culture.

ACADEMIC AND SCHOOL GEOGRAPHY

Though the National Geographic Society quickly became a popular and national institution, its origins lie in the small community of scientists working in Washington, D.C. after the Civil War. The Society's first generation of members witnessed, and in some cases facilitated, the migration of science from the federal government into the new and reorganized universities around the nation. Thus the Society is important not just for its twentieth-century popular incarnation, but also for its relationship to academic geography in the late nineteenth century. The academic study of geography in nineteenth-century America was largely understood to be a tool of exploration and conquest, an aid to the surveying, mapping, and development of the West. But by 1887 Halford J. Mackinder, a pioneer in British geography, proclaimed that "a Stanley can never again reveal a Congo to the delighted world," and a few years later Frederick Jackson Turner cited the end of a continuous frontier line in the American West as an epochal moment in the nation's history.[10] Though these statements by no means signaled the end of exploration, the late nineteenth century was indeed a turning point for geography as a science. As a relative latecomer to the newly organized sciences of the late nineteenth century, geography was forced to define itself in an era of ever increasing academic specialization. In the university context we find clear articulations of geography's purpose made by self-conscious professionals who were determined to create a scientific rationale for their study. Struggling to secure a place in the university, American geographers began to distance themselves from existing geographical and exploration societies such as the National Geographic Society, which they now considered excessively broad and amateur. Instead they concentrated on building a specialized university science that might earn them professional standing. Academic geographers were preoccupied with their institutional legitimacy throughout this period, and this preoccupation in turn shaped their discipline. As David N. Livingstone has argued, geographers came to defend their niche as the special relationship between humans and their environment, and this concept led them to experiment with a range of explanatory frameworks, including variants of Lamarckism and Social Darwinism.[11] Their desire for academic legitimacy prompted them to cling to these frameworks long after they had been discredited by other disciplines. Geography's need for legitimacy—both intellectual and institutional—directly influenced its development within the university from the late nineteenth to the early twentieth century.

Geographers also provided intellectual coherence and justification for the nation's commercial and political expansion. The link between academic geography and expansionism here is complex.[12] American geographers were occasionally called to serve the state—such as in the United States Geological Survey or the Paris Peace Conference—but more important were the indirect links between geographers and the knowledge they created. Many American geographers applied evolutionary theory to environmental circumstance in order to integrate the human and natural worlds. By concentrating on the ways humans acted relative to their environment, academic geography demonstrated the need for the wisdom, benevolence, and efficiency of American foreign intervention. This intellectual model shaped school geography throughout the first half of the twentieth century. But if we claim that American geography implicitly supported American expansionism, we must also recognize the reciprocal nature of this relationship. As Livingstone has argued in the British context, "imperialism no more provides the context for geography, than geography provides the context for imperialism."[13] Just as the National Geographic Society's relationship to the Spanish-American War suggests the union of science and politics, so too did America's interests abroad help shape the discipline at the turn of the century. As chapter 4 argues, it was in this period—from the 1890s to the eve of World War I—that university geography took on some of its basic intellectual and institutional characteristics. Beyond this chapter, academic geography is generally treated only insofar as it influences other, more general developments, such as the development of American school geography.

Academic geographers took great pains to emancipate school geography from the drill-oriented form that prevailed in the nineteenth century and to transform it into a dynamic subject appropriate to the goals of modern education. In the late nineteenth century, a geography tutor in a small southern town commanding his students to "Bound Beloochistan" expected them to name the regions surrounding what is now southwest Pakistan: Persia, Afghanistan, British India, and the Arabian Sea.[14] These pupils encountered the world as a curious, exotic array of states, races, and cultures ranging from enlightened to barbarous, from Christian to heathen, from prosperous to destitute. They memorized facts and figures about a world peopled by cunning Chinese, militaristic French, orderly British, and indolent Spanish, transmitted through legends and the discoveries of explorers both past and present. They read textbooks that organized the world into timeless and immutable hierarchies of civilization, closely correlated to hierarchies of race, which presented the world as a distant, alien, and rigidly defined spectacle. Geography was also, more often than not, a lens through which the wonders of the world could be discovered. To this end, contemporary geography textbooks dwelled on the awesome, dramatic, and majestic aspects of the natural landscape—the tallest mountains, the longest rivers, and the climatic divisions of the world—as part of a divine plan, and were slow to incorporate the dynamic implications of Darwinian evolution in any thorough sense. Because it considered all of the natural world, nineteenth-century school geography encompassed many other subjects, including most importantly history and geology.

By the late 1890s, however, the university professionals discussed above had refashioned geography into a systematic, scientific study, for both their own prestige and the survival of the subject in the secondary school curriculum. Their efforts brought the study of the physical world into an evolutionary framework and into the modern system of education. Modern geography centered around process rather than place. By extension, modern school geography was no longer a discovery of Christian design but a way to identify and negotiate the nation's role in the world. Descriptions

of the natural world quickly lost their teleological overtones, yet the dynamic form of geography remained a deeply evaluative and moral subject
lesson. Political imperatives at home and abroad at the turn of the century
also brought new goals to geographical education. With the Spanish-
American War the United States raised its commitment to world commerce, and geographers showed their enthusiasm for this new internationalism by writing texts that presented the world not simply in terms of the
relationship of race to region and climate, but rather in terms of the relationship of nations and colonies to resources. In this regard commercial
geography fulfilled the utilitarian demands of the progressive curriculum.
While nineteenth-century school geography had framed nature as a symbol of divine providence, twentieth-century school geography understood
nature as, above all, an opportunity. By focusing on the extraction of natural resources, it also connected humans to their environment while appealing to a set of national priorities contingent upon trade. Commercial geography undermined the world of racial and natural exoticism by describing
a more immediate international community where—despite differences—
all humans and nations contributed to trade through the conquest of the
land. Correspondingly, civilization began to be judged increasingly by reference to commerce. School geography developed according to this formula, adjusted over time, from the early century through the 1930s. Though
an independent subject through World War I, it was eventually integrated
into the social studies, modified as well by interest in the economic dimension of the subject.

In the 1930s war again turned public attention abroad, and quickly
transformed the subject's character in schools. The arrival of the air age and
the direct threat of war forced textbooks to bring geography into a far more
visual, and explicitly spatial, dimension. Geography began to focus more
directly on the relations between states in a world of total war. The war and
the air age brought nations closer together across the seas and over the
North Pole, raising both hopes for the future of international trade and
communication and fears about proximity to the enemy. The cold war furthered this politicization of space, and encouraged a cult of geopolitics that
infused texts throughout the 1940s. Students now learned about the implications of the air age for the political relationships between states. While
most prewar texts were titled "Commercial" or "Economic" Geography, by
the 1940s students studied "Global," "Air Age," or "Political" Geography.
This new world order was threatened by differences not of race but of political and economic ideology. In this regard geography's history in American schools has been fluid. Within the curriculum geography moved from

a study subsuming both history and physical science, to one geared toward commerce, to one focused on geopolitics. As chapters 5 and 6 argue, school geography throughout this period narrated a world filtered by political and economic imperatives, tailored to the character and intensity of the nation's commitments abroad.

These loosely defined traditions—maps and atlases, the *National Geographic*, academic and school geography—produced the dominant narratives and graphic constructions of geography. Though they often conflicted and competed with one another, close study of the three together reveals two overarching points. First, geographical knowledge was used to describe and contain a world in which the American presence was rapidly expanding. As an attempt to identify and define the nation it quite often buttressed the designs of the state, even though these efforts generally came not from government sources but from individuals and private organizations. For instance, the Spanish-American War sparked a reorientation in school geographies, atlases, and the ever more seductive work of the National Geographic Society. Cartography companies added large maps of the Philippines and Puerto Rico to the world atlas, revising the narrative content as well to include enthusiastic descriptions of the landscape, people, and resource wealth of the new territories. This was a world organized around commercial potential rather than racial difference, one where the place of the United States was actively touted rather than isolated and separated. In these and other ways the accumulation, organization, and dissemination of geographical information has often been a struggle to present and legitimate the nation and its goals.

Yet the history of geographical knowledge in American culture cannot be understood solely as a function of state power. For example, a cartographic revolution took place—unexpectedly—when William Rand and Andrew McNally began a printing business that eventually brought cheap, mass-produced maps to the American public. The advances of print influenced not just the form and content but also the production, distribution, and culture of American cartography. Furthermore, important innovations that ought to have affected maps were often suppressed. The advent of flight in the early twentieth century might have dramatically affected map perspectives and projections, but this reconsideration was delayed because the nation's largest mapmaking firms were slow to adopt new mapmaking techniques and reluctant to challenge existing cartographic conventions. These choices, like the cartographic industry's decision to rely for so long on the Mercator projection, indicate that the history of geographic knowledge is complex and subject to multiple, sometimes contradictory influences. And often more important than instances of change are those conventions that survive unchallenged over time, eventually becoming accepted as reflexive truths. This study explores how geographical knowledge works to create and entrench particular ideas about the world and the place of the United States within it.

PART 1

Making Geography Modern

2

Maps for the Masses
1880–1900

World atlases were scarce in early nineteenth-century America. For cultural reasons, the nature of cartographic curiosity among Americans was more national than international. And for technological reasons, mapmaking was a laborious and costly process undertaken by only a handful of companies in the American northeast. While print culture spread dramatically in the early part of the century, mapmaking techniques continued to produce atlases that—though aesthetically impressive—remained prohibitively expensive. By the 1880s, however, American cartography had evolved from an elite craft evoking artistic and utilitarian sensibilities to a mass industry producing maps as quickly and cheaply as possible. New technologies, organizational modes, and economic motives facilitated the spread of inexpensive atlases through American culture, a trend that made atlases a common fixture in urban parlors and rural schoolrooms alike. Yet declining prices and rising availibility could only encourage—never guarantee—the national appetite for atlases. Critical to sustaining this demand was the explosive growth of railroads, national consolidation after the Civil War, and a widening vision of the world that climaxed in 1898. In fact, the reorganization of cartography from a craft into a commodity was especially important because it arrived on the eve of a war that increased the nation's territorial reach and transformed the scope of the world atlas. This chapter examines the design, production, marketing, and consumption of the atlas from the 1870s to the end of the nineteenth century in order to demonstrate how cartographic companies helped Americans make sense of a world that was far too complicated to comprehend on its own.

THE ORIGINS OF THE MODERN WORLD ATLAS

Situating the modern world atlas requires a general familiarity with the geographic print culture of the early nineteenth century, one of the most important features of which was the regional gazetteer. In essence, the gazetteer was a reference tool, a dictionary that catalogued and indexed a particular region, and occasionally the nation as a whole. Though the genre dates back to the early eighteenth century, regional gazetteers reached the height of their popularity in America during the early national and antebellum periods. In part, the rise of the gazetteer stemmed from the adaptation of the genre to the needs of the market. As Patricia Cline Cohen has observed, beginning in the 1790s the gazetteer moved to reject the "casual and imprecise" facts of its antecedents in favor of a more statistically oriented form that was geared to following trends and documenting aggregate change. By supplying statistics related to manufacturing and agricultural output, transportation infrastructure, and population distribution, gazetteers became manuals of economic geography designed to chronicle progress from one era to the next.[1]

Because they were concerned with the realm of human activity, gazetteers detailed the physical geography of a region primarily as it related to the economy, such as its proximity to waterways, its agricultural conditions, its industrial strengths, and its commercial potential. In both design and format gazetteers profiled a region in terms of its elements rather than as a unified landscape, and in this sense they were part of the larger nineteenth-century trend toward commodifying nature.[2] Gazetteers took the landscape, measured it, broke it down, packaged it, and sold it, and if the genre was understood as "geographical," it was a model of geography that referenced discrete locations rather than relationships. Though the relevance of the gazetteer declined with the rise of industrial capitalism, this reference quality—an interest in locations rather than relationships— influenced what came later. In fact, the late nineteenth-century atlas was also commonly termed an "atlas and gazetteer." As we will see, the atlas incorporated this interest in quantification and statistics even though the genre's broader and more diffuse goals often made such information anachronistic. The world atlas of the late nineteenth century was also influenced by early nineteenth-century atlases, which—like gazetteers—often focused on the county. At the peak of their popularity—just after gazetteers, from 1850 to 1880—the county atlases directly addressed the creation and growth of the West. They generally included a map of the state or county in question, maps of landholdings in the individual townships with descriptions of the local region and its history, and biographical sketches of its more prominent citizens.[3] In this regard the county atlases grew out of, and were close cousins to, gazetteers.

Just as popular were the geographical readers of the early nineteenth century, which linked geography with the concerns of nationhood. The relationship between geography, maps, and early American nationalism has been explored by Martin Brückner, who argues that the wide diffusion of geography texts and readers in the early republic connected "the Vermont lawyer and the Connecticut farmer, the New England statesman and the southern schoolgirl."[4] Take for instance Jedediah Morse's *Geography Made Easy*, an instant bestseller in 1784 and commonly read through the 1820s, surpassed in popularity only by the Bible and Noah Webster's spellers. While Webster modified the English language to help Americans gain intellectual independence from Britain, Morse used geography to recast American understandings of their home territory. As Morse explained to his readers, while in the past "our youth have been educated, rather as subjects of the British king, than as citizens of a free and independent Republic. . . . the scene is now changed . . . particularly to that of the Geography of our own country."[5] Webster would lay the linguistic foundation for a true and pure national identity, and Morse would unite Americans by creating a common territorial and topographic basis for nationhood.

The work of Benedict Anderson is central to understanding the place of geographical texts in the creation of nationalism. For Anderson the map anticipates the nation by concretizing, or at the very least elaborating, the "imagined community" of distant fellow citizens. For example, Anderson cites the importance of what he terms the "map-as-logo," an image of the country abstracted into an iconographic representation, which gives citizens an instantly recognizable symbol of a "home" that would otherwise be unfamiliar.[6] And, in fact, each of Morse's two bestselling readers, *Geography Made Easy* and *American Universal Geography*, contained a map of the new nation, a highly accessible symbol that unified a disparate citizenry by speaking to all Americans on a visual level. Furthermore, Brückner cites the presence of national maps and other tools of geography—such as textbooks, compasses, and charts—in contemporary portraits, used to signify national pride as well as social status and education of the new middle class. Through these examples Brückner argues that maps and other geographic symbols of the young nation cemented the more abstract ideologies of nationalism in later decades. More essentially, Brückner writes, the national map became a preeminent and reflexive image that was nearly impossible to either transcend or reject; this leads him eventually to ask whether there can even *be* a nation without a map.

The structural power of cartography to create both physical and cultural space is also evoked by the early nineteenth-century debate over the prime meridian. As Matthew Edney has argued, cartography was understood in the late Enlightenment as an "archival corpus" that would rationally organize the collective knowledge of a given territory. Cartography relied upon the graticule—the framework of meridians and parallels—as a way to plot locations and other features of the landscape accurately. But to reconcile the competing and overlapping maps of a given territory, one had first to create and accept a common language, or point of reference, and for cartography that reference point was a prime meridian, the line against which all lines of longitude were measured. The conventional prime meridian prior to the nineteenth century had been drawn through the Canary Islands, considered a "natural" choice because it bisected the Atlantic rather than a land mass, and neatly separated the Eastern and Western Hemispheres. But such a location was a product of human deliberation, just as the later proposals for meridians through Paris and Greenwich were justified by the existence of observatories in those cities. Similarly, the decision to have a universal prime meridian stemmed from an Enlightenment concern with universal modes of knowledge and uniform standards of measurement, reflected by the desire to collect and rationalize all information into a single body of knowledge.

Against this universal meridian came the movement to establish a national meridian in the United States, based on the argument that the calculation of longitude from a distant, foreign meridian signaled a colonial, dependent relationship. An American amateur astronomer, William Lambert, proposed that the meridian run through Washington, D.C., both a cultural and a physical symbol of the nation's ascendancy and independence. The attempt to establish a national meridian illustrates Edney's general claim that the relationship between the map and the territory is an intimate one. The prime meridian defines and charts physical space, yet it was conceived not just as a measurement of the physical world but also as an element of human society, even a tool of political ideology. In Edney's words, "As lines established by human agency to define longitude, otherwise an attribute of purely physical space, prime meridians were held to mediate between the concrete extent of the territorial state and the abstract extent of the cultural nation."[7]

Just as Brückner sees the power of the American map to become a reflexive image of national power, Edney sees the prime meridian as a marker of physical space that takes on cultural and political meaning. The debate over the universal prime meridian suggests that "social space was seen as being structured in the same manner as physical space, which is to say that representations of social space were constructed through maps and mapmaking." By taking this crucial step, Edney writes, "cultural boundaries became as real and as untraversable as a physical cliff; the gradations of the natural world became as hard and fast as property lines."[8] Somewhere in the process, the map had been taken for the territory, blurring the line between physical and social space. Both Edney and Brückner suggest that cartography was a critical form of national expression in the early republic, a representation of territorial union that undergirded a more abstract understanding of political union.

National cartography was also essential to the growth of trade and the creation of wealth in the new republic. As Edney argues, "without the creation of property through the mapping and knowing of space, the existing social order could neither be confirmed nor extended into virgin territories."[9] This made surveys of the land not just useful but essential, and fueled the disproportionate cartographic attention given the United States. This parochialism also has more mundane explanations. Because American-made world atlases could not easily compete with their European counterparts, the work of this period tended to concentrate on the United States relative to the wider world.

Scholars have characterized this period as the "golden age" of American cartography, initiated by the publication in 1795 of Matthew Carey's *American Atlas*. Nearly all of the atlases of this era were printed with copperplates, a method that produced maps far more durable than those of later years. Also known as the intaglio process, copperplate engraving allowed fine, delicate lines that were easily maintained and updated; subtle and variable area tones; and small, neat lettering. All of these qualities created a map considered aesthetically superior to those produced with woodcuts or other early methods (fig. 2.1). Yet this process also required great skill and painstaking attention to detail, and produced a limited number of maps priced well beyond the reach of most potential consumers.[10]

Lithography—a more versatile method that produced lines that were delicate but not as clean as those created with copperplates—became increasingly popular by the 1840s. The rise of lithography coincided with a period of tremendous internal growth in the United States, particularly in canals and rails, developments that in turn generated a demand for maps. But many mapmakers, such as Samuel Augustus Mitchell and J. H. Colton, continued to use copperplate engraving through the 1840s and 1850s, and both this and the lithographic method kept atlases relatively expensive. On the eve of the Civil War, then, access to atlases was still limited to the few who could afford them.[11] The war itself exposed many Americans—especially northerners—to maps on a daily basis. These were not the products of skilled cartographers striving for topographic accuracy but rather of journalists and artists hoping to convey a sense of battlefields and their general geographic situation. Frequently pirated without acknowledgment of authorship, newspaper maps began a tradition of popular mapping during wartime that would grow exponentially in the twentieth century.[12] The Civil War also influenced map production in more general ways. By introducing an era of industrialization and urbanization, the war catalyzed the growth of railroads linking East and West, North and South, urban and rural. Movement into the West demanded knowledge of the land. Prior to the Civil War, federal surveys had been limited to the coasts and later to the public domain, such as John Fremont's expeditions in the 1840s for the Corps of Topographical Engineers.[13] After the war, the need for accurate surveys of the interior increased, yet the federal government was slow to respond in any formal, systematic way. The immense task of mapping the West clashed directly with the strong tradition of nineteenth-century

Figure 2.1   Detail of a map produced through copperplate engraving in the mid-nineteenth century. A costly and time-consuming process, copperplate engraving produced maps characterized by varied size and style of lettering and a degree of selectivity in determining what appeared on the map. Courtesy of the Harvard Map Collection.

federalism. The disorganized nature of these western surveys resulted in extensive duplication and intense rivalry for funding and prestige. Consequently, in 1879 Congress, with the help of the National Academy of Sciences, founded the United States Geological Survey, and charged it with studying the geological structure and economic resources of the public domain.[14] By the 1880s and 1890s, the USGS, the Coast and Geodetic Survey, the Hydrographic Office, and the Army Corps of Engineers were each producing and selling inexpensive topographical maps that familiarized the public with mapping and eventually stimulated a demand for atlases.

The migration westward also brought Chicago into the limelight for its ability to serve these new rural markets better than Baltimore, Philadelphia, New York, or Boston. Burgeoning new rural and town populations showed steady interest in buying books, and by the 1870s Chicago had established its reputation as a bookselling town. Soon thereafter publishing houses began to appear, some of which recognized that alongside this growing market for books was a budding interest in domestic maps.[15] Chicago's late entry into mapmaking hindered it from competing effectively with the established cartographic companies of the northeast, yet the city was crucial in satisfying a qualitatively different—but rapidly growing—market for maps. These new consumers—less interested in the "fanciful appeal," "aesthetic design," or "flawless execution" offered by copper engraving—needed maps of the railroads and surrounding towns and of county divisions, maps that emphasized political and economic information even at the expense of artistic detail.[16] Above all, this information had to be affordable, even cheap. The price of copper engraving had never allowed for mass production, and some firms had even consciously rejected newer techniques in order to maintain their traditional—though limited—market in both Europe and America. Thus, the potentially wide demand for inexpensive maps was left to be met first through improvements in lithography, and then more consequentially through wax engraving.[17]

Arthur Robinson, a cartographic historian, has suggested that these changing print technologies influenced American cartography even more than the evolution of surveying techniques, and the impact of wax engraving bears this out. During its 110-year life span, from 1840 to 1950, wax engraving—or cerography—was responsible for nearly 75 percent of all regional, national, and world atlases on the American commercial market; between 1910 and 1920 that figure reached 95 percent. As one mapmaker commented, wax engraving "has done more than any other one thing in America to put the map to work among millions of people."[18] It is no exaggeration to say that the widespread adoption of wax engraving after the Civil War shaped the production style and scale of American maps for decades. Any understanding of how Americans encountered their world through maps in this period must recognize the implications of this technology.

Wax engraving was thoroughly American. Created by an American, adopted only by American companies, and building a style that would shape the American map, the technology goes a long way toward explaining the survival of some of the largest map companies founded in this period. In part, wax engraving dominated mapmaking for so long because the American cartographic community was relatively small. As a print technique wax engraving involved no formal training; the trade was acquired primarily through apprenticeship in the nineteenth century and through in-house training in the twentieth. Frequent movement of talent between firms helped spread trade secrets and conventions that in turn homogenized the production of American maps. After twenty years at Rand McNally, for instance, Caleb Hammond left to found the Hammond Map Company — soon to become a major producer — taking with him his accumulated experience. Wax engraving was also the first map-printing technique developed fully within the era of industrialization, involving a kind of assembly line production. As such, it represented a dramatic departure from traditional cartography. In the copperplate process, the cartographer and engraver worked closely to produce the map; with wax engraving, cartographers were often replaced by draftsmen who had considerably less experience and who became increasingly specialized workers. This change refashioned cartography from a craft into an industry. The aesthetics of a map were now overshadowed by the imperatives of mass production, and the premise of nineteenth-century cartography — aesthetics dictating price — had been undermined by the 1880s through a process of commodification. This change would have consequences for the role of maps in American culture.[19]

RAND MCNALLY, GEORGE CRAM, AND THE QUEST FOR INFORMATION

Though Rand McNally was not the first firm to use wax engraving, it was the most successful. In 1868, William Rand and Andrew McNally formed a partnership, purchased the *Chicago Tribune*'s job printing department, and began producing tickets and other materials related to the explosive growth of the railroads. Within a few years, the firm was advertising that, "having access to the best Military and other Government maps . . . [it is] prepared to furnish drawings of any size and scale."[20] By the 1880s Rand McNally's name was synonymous with maps, which made up about 40 percent of the company's revenues. An 1887 Chicago business guide boasted that

Wherever, on the American continent, railways are known or
books are read, the name of Rand, McNally & Co. is familiar . . .
their atlases, maps, and directories are seen in the business offices
of every city and town. . . . [It is] unrivaled in the world at the
present day. . . . In short, there are few establishments in the world
whose names are so widely or so favorably known as that of Rand
McNally & Co.[21]

Though certainly more promotion than fact, the excerpt reflects a growing recognition of the company's maps, at least in the Midwest. The company grew steadily in the 1870s and 1880s, aided immensely by Andrew
McNally's position in Chicago's business and cultural community. Though
the depression of the 1890s savagely cut the company's net profits, for
McNally himself it was a crowning decade. He was appointed to the Board
of Directors of the World's Columbian Exposition, while his company was
commissioned to print the official guides and maps to the fair itself. An
Irish immigrant, McNally became a kind of Horatio Alger figure in Chicago, accorded the honor of opening the gates of the Exposition's White
City. Admired among his peers as the "representative printer of Chicago,"
at the end of the century McNally bought out Rand's share in a company
that now had six divisions: tickets, railroad printing, map publications,
bank publications, trade books, and educational textbooks. Half a century
later, Andrew McNally III attributed the company's early success to one
thing above all others—the decision to produce maps with wax engraving
technology.[22]

At the time, Rand McNally's only serious rival in wax engraved maps
was George F. Cram & Company, started by a nephew of a cartographer who
had long worked for Sidney Edwards Morse, son of Jedediah Morse. Cram
established his own firm in 1869 and began publishing atlases immediately,
gradually building a national market by using wax engraving and materials that were even less expensive than Rand McNally's.[23] But despite its
rapid proliferation in the United States, wax engraving never caught on
in Europe, a fact that widened the gulf between American and European
cartography. European cartographers considered wax engraving an inferior
method. Moreover, because they were still selling to a more limited market, the mass-production capacities of wax engraving held little attraction
for them.[24] Historians of cartography have been particularly critical of
cerography for transforming an age-old craft into an industry where technology—rather than artistry—shaped the final product. One has even described the era of wax engraving a kind of cartographic dark age where "bad
maps" dominated the market. Michael Conzen has referred to the "crudity"

of wax engraving, while David Woodward has claimed that the technique broke the aesthetic tradition and damaged the nation's "cartographic consciousness." Woodward speculated that the monotonous, over-lettered style of this method dulled the population's sensitivity to the use and beauty of maps like those produced in Europe. To illustrate the pernicious reach of wax engraving, he pointed out that even maps produced with other techniques imitated its unappealing style because many Americans had come to identify it as an ideal cartographic type, what a map *ought* to look like. But in order to distinguish what Americans wanted in a map from what they were offered, we need to examine patterns of both production and consumption.[25]

There is no question that wax engraving brought a different kind of map into circulation. Crammed full of type, these new maps located as many towns as possible rather than selecting out the largest or most important (fig. 2.2). Compare this map detail to that found in figure 2.1. Though both cover the same area of southwest Washington, the latter deemphasizes topographic detail in favor of more place names, a clearer identification of county divisions, and greater attention to the growth of towns along the railroad. Notice also the more uniform style—and smaller size—of lettering in the map created through wax engraving. There are a few explanations for this. Woodward has suggested that wax engraving often left decisions about what to include on the map up to the engraver or stamper, who usually had little or no cartographic experience. Engravers were often given only vague direction about what to include on the map, and since lettering was done not by hand but with printer's type, many locations could be added, even if that meant moving names in order to include them all. Further, wax engraving was poorly suited to reproducing halftones, and thus discouraged the representation of graded features on a map. Instead, wax engraving relied on hatching or contour lines to indicate topographic detail.[26]

But technology alone cannot explain the appearance of these maps, and something about them must have satisfied the consumer. A close look at Rand McNally's advertising suggests that the desire for information, rather than artistry, was central in the late nineteenth century. American-made maps and atlases were rarely advertised as objects of beauty, or even products of advanced cartographic techniques. Instead, marketing emphasized accuracy and an unmatched number of locations. An 1886 ad for *The Indexed Atlas of the World* described a range of information such as comparative national debt, population, taxation, wealth, density of population, and religious divisions of the world. Like an almanac, the atlas was one "of human affairs" aimed at those wishing to be better informed as well as those

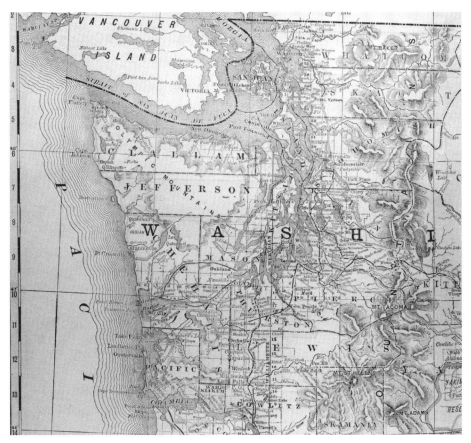

Figure 2.2   This detail of a Rand McNally map was produced with the late nineteenth-century American innovation of wax engraving. Notice the appearance of this map detail relative to that in figure 2.1.

needing a reference work. An 1893 issue of *Godey's Lady Book* advertised the same atlas by quoting testimonials to its accuracy and its unsurpassed amount of information, including "400,000 places, diagrams under 150 separate headings, with 7,000 single items."[27] These advertisements rarely touted the aesthetic beauty of the atlases—in this respect they simply could not rival their European counterparts—and some even addressed the perception that Rand McNally maps were inferior. But they *could* compete in terms of price and amount of information included, so both became strong selling points: an ad in *Youth's Companion* cheerfully announced that consumers would no longer be forced into buying expensive British atlases with limited coverage of the United States.[28] Many of these maps were created with the businessman in mind, one in charge of distribution and sales who needed to know more about town size, name, location, and rail systems than about topographic or geographic relationships. The American map industry was primarily driven not by aesthetics but by a utilitarian need to map terrain in the West and the expansion of railroads and markets nationwide.[29]

This relationship between American cartography and the growth of railroads is critical. Rand McNally directly linked its products to the growth of the rails by selling books and atlases in trains and stations, and by distributing complimentary copies to promote a particular rail network the company had mapped. The more people knew about the railroad, the company reasoned, the more use they would have for maps of it, and who better to promote a railroad's route than a map company? In fact, there are even instances where mapmakers manipulated their maps in order to improve the appearance of a particular railroad's reach.[30] The company also distributed atlases through bookstores and through direct subscriptions that allowed consumers to acquire an atlas with lower initial outlays and payments over time. None of the more established map companies of the northeast were then selling by subscription, further fueling Rand McNally's new popularity with the broad middle class. An even more common method of distribution was through newspapers, which purchased the atlases from the company and then sold them to readers—either in parts or as a whole—often as an incentive to subscribe to the newspaper.[31] Like Rand McNally, the George F. Cram Company produced atlases to fit a variety of incomes and needs at the turn of the century.

READING THE ATLAS

At first glance, late-century atlases appear to be arbitrary collections of information covering everything from cotton exports to the world's longest rivers. In this respect they are the direct descendants of gazetteers.[32] This abundance of "facts" and locations on the map and within the atlas suggests a veneer of encyclopedic objectivity, yet the volume of information should not obscure the presence of motive or rhetorical form. Though frequently idiosyncratic in their selection and presentation of information, atlases were deliberately constructed texts that contained particular kinds of maps, graphs, tables, pictures, and narrative, all of which purported to explain, and contain, the world and its peoples. Even the sequence of the maps suggested a narrative: maps of the United States began in the northeast then fanned south and west, while coverage of Europe invariably began with the British Isles before moving east to the continent. In both

cases the point of origin was the Anglo-American community that bor-
dered the North Atlantic. Just as striking is the cartographic focus on the
United States in these early world atlases. Rand McNally's *New Household
Atlas of the World* (1885) devoted two pages to a map of Alabama, while the
entire continent of Africa was reduced to a single page. In all, 122 of the
136 pages — roughly 80 percent — were devoted to the states of the union.
In the 1890 edition of Cram's *Unrivaled Atlas of the World*, 136 of the 181
pages of maps — 77 percent — covered American states and cities. Though
designated "world atlases," maps of the United States constituted the great
majority of the total at late century.

Compare this ratio to that of the European atlases. Germany's popular
Stieler's *Atlas of Modern Geography* devoted just 5 to 7 percent of the maps
in a late nineteenth-century edition to German regions. John Bartholo-
mew and Son, the leading Scottish cartographer, generally reserved about
10 percent of its maps for Scotland, and about 18 percent for the United
Kingdom. Even allowing for large differences in land area — Scotland and
Germany are far smaller than the United States — this comparison suggests that the disproportionate emphasis on the home country was uniquely
American. Qualifications to this generalization do exist, however. In Austrian school atlases around 1870, European maps constituted more than
80 percent of the total. Though not a parallel comparison, because a national United States bias is not the same as a continental European bias,
this emphasis does reflect a similar local prejudice.[33] Overall, the late
nineteenth-century world atlases produced in the United States were overwhelmingly national. In fact, the very form of the atlas separated the
United States from the rest of the world. Within this parochialism, we find
an even stronger state and regional bias, and as late as 1908 the Cram
Company was appealing to state loyalties with atlases such as the *Superior
Reference Atlas of Michigan and the World* or the *Superior Reference Atlas of
California, Nevada, and the World*.

Despite their focus on the United States, these atlases did include at
least one map of each continent, though subcontinental coverage — the
number of maps given to areas within a continent — was highly uneven and
reflects relative interest.[34] Another measure of relative emphasis in atlases
is the scale of coverage. How comprehensively a region is mapped indicates
what is known of the region and also its perceived importance. That central Africa was virtually ignored until the early twentieth century reflects
both the rough knowledge of the physical landscape and its relative insignificance to Americans. There are certain problems with such a method
of comparison, especially the danger of misinterpreting this difference: often scale distortion can be explained by factors unrelated to cultural bias

or political ideology. That Spain is generally found on a map of the entire Iberian Peninsula could be attributable to its geographic position and shape or to a perception that the nation of Spain is somehow subordinate to the peninsula. Thus the convention of mapping Iberia as a whole cannot be said to reflect simply a lack of respect for Spanish political integrity. Yet the massive island of Madagascar—approximately the same size as Iberia— was never mapped independently, even though both its size and its geographic integrity made such a cartographic division logical.

So while some cartographic conventions can be explained by geographic circumstance, many more have simply become "naturalized" through repetition over time. Consider the tendency to focus on western Europe, an emphasis that can be explained by American political, ethnic, and historical ties to the region. Another explanation for this regional emphasis would be that the political and economic strength of European nations earned them relative prominence in the atlas. But these conventions are historical, and as such we ought to be aware of their power. In fact, almost any atlas since the late nineteenth century legitimates European and American nationalism. The "continent" of Europe has generally been divided along national borders, which over time became naturalized despite their fluidity. This emphasis on western Europe is reflected by the hemispheric maps that opened most contemporary atlases. In a Cram atlas of 1890, the Eastern Hemisphere is laid out with Europe at the center rather than the far west (left).[35] Subcontinental coverage also typically favored western Europe: the 1883 edition of Cram's Unrivaled Atlas printed subcontinental maps of European nations alone, while in Rand McNally's 1885 Household Atlas only the British Isles, Germany, and Scandinavia merited separate maps. Soon, however, more intense coverage of areas outside western Europe began to appear, usually where colonial interests were strongest or in areas that figured prominently in the narration of western history. Rand McNally's Pocket Atlas of 1887 included regional maps of southeastern Europe, north and northwest Africa, southern Africa, and Persia, Afghanistan, and Baluchistan (now southwest Pakistan), while South America was arbitrarily divided into three lateral regions.

One of Rand McNally's most ambitious efforts of this period—the New Family Atlas of the World—mapped each European nation independently while giving South America, Asia, and Africa fewer subcontinental maps. The atlas outlined northern and southern Africa in regional maps while omitting central and sub-Saharan Africa altogether. In Asia, only Japan, China, and the Ottoman Empire were given regional maps, as was the region of Persia, Afghanistan, and Baluchistan. This organization elevated the European nation while ignoring other national boundaries in favor of larger regions, even where clear and strong national identities existed. For example, in Cram's *Unrivaled Atlas* (1890) most western and northern European nations were mapped separately, while Latin America, made up of politically independent nations by the 1820s, was mapped as a continent. Well into the twentieth century South America would be mapped and identified according to regional rather than national divisions, irrespective of the continent's political history. In the process, the physical continent became the primary unit of identification, a decision that silently but effectively naturalized South America and placed it outside of history.[36]

The use of scale also quietly emphasized western Europe. The scale of European national maps in Cram's *Unrivaled Atlas* became as detailed as one inch representing seventeen miles for Switzerland and as broad as one inch equaling three hundred miles for the East Indies, or one inch representing 135 miles for Egypt, Arabia, Upper Nubia, and Abyssinia. Furthermore, many areas were not mapped at all except as part of the continental map, such as the Sahara and central Africa. Though much of this can be explained by the extant knowledge of these regions—for example, cartographic data for sub-Saharan Africa was still relatively scant in the 1880s and 1890s—even more important was the fact that Rand McNally and Cram often took cues from more established mapmakers, generally European.

In general, these late nineteenth-century atlas maps featured political boundaries, location and transportation information, and some topographical information such as waterways and large mountain ranges. National boundaries were prominently outlined, as were railroad lines, rivers, and lakes. But most noticeable was the tendency toward crowding information, and dense lettering quickly became the rather notorious trademark of American maps. Ironically, it was the seemingly democratic practice of including as many towns as possible—a practice facilitated by the methods of wax engraving—that inadvertently transformed the nature of American maps. By identifying as many towns as possible, regardless of size, the maps gave little indication of population distribution, instead suggesting that all areas were equally settled (fig. 2.3). These maps encouraged readers to identify discrete locations rather than to explore relationships, a fact confirmed by the ever more comprehensive indexes at the end of these atlases that listed virtually any town or village. This reference quality of the atlases was their strongest selling point, but in the process of compiling and presenting all that information topographical relationships—such as the contours of the land and relative elevations—were sacrificed. By contrast,

Figure 2.3 With so many place names, this map detail (Rand McNally, 1890) gives the reader little indication of the relative population distribution in the state of Missouri. The railroad and the town are treated as defining features of the landscape.

most European maps of this period included fewer place names in order to make room for other kinds of information, a practice that ultimately allowed for greater demographic precision.[37]

The noncartographic pages of the American-made atlases—filled with graphs, charts, and text—gave more attention to the larger world than did the maps themselves. In part this reflects the availability of national census records and other sources. As a result we find an emphasis on "quantifiable" information such as industrial production, ethnic breakdown, and religious association. Though apparently idiosyncratic, the information generally related to the human rather than to the natural world. These atlases framed the world as a racial hierarchy by highlighting the unified relationship between race, climate, and "progress," and in the process created an ethnographic world that functioned according to certain laws. This approach was reinforced by the descriptive text of the atlases, which described the nations of the world, their races, social customs, and "the part they have played in the progress of civilization." As a result of cartographers' treating the nation as a product of both the physical environment and the racial constitution of its inhabitants, race and nation—sometimes race and continent—became conflated. Thus the text characterized national populations through social qualities such as indolence, gaiety, or aggression. Rand McNally's *Pictorial Atlas* of 1898 began its narrative text with a description of North Americans as imbued with an ardent love of liberty that was shaped by both biology and environment. The innate "genius of Anglo-Saxon industry" was influential, but so was the natural wealth of the American environment and the "fostering care of a wise and liberal government." This quality was potentially transmuted to the more recent immigrants and their American-born offspring, and gradual assimilation of immigrants would allow "a new people [to] arise, still distinctive in its nationality, yet composite and comprehensive in character, and in its relations to the state and to society combining somewhat of European conservatism with republican individuality and force." "Conservatism" was not simply a political characteristic, or even a personality trait, but an innately European quality. Through this language nations and races became conflated further, and characterized through qualities such as "industrious" or "lazy." Thus the destiny of America was to be the model of democracy for the rest of the world, blessed as the people were with not only an abundant environment but also an innate racial and political genius and superiority.[38]

By contrast, the history and future destiny of Mexico was one not of exceptionalism but of "tragic interest." In the Mexicans was found not the "ceaseless energy which in the United States has attained such marvelous results" but rather a natural lethargy, attributable to both climate and heredity. Though the Mexican was "quick in his perceptions, enthusiastic by nature, and in his general deportment dignified and courtly," these mannerisms were better suited to ancient Spain, and in modern Mexico all too often created "a public highwayman, a professional gambler, or a political revolutionist of doubtful motives." South America, too, had the potential for civilization, apparent in the independence movements that had stirred in the people the same love of freedom so important to Americans. But the "revolutionary character of the people" would hinder any smooth or stable transition to full democracy. Both South America and Mexico had inherited too many Spanish traits, which included gentility and refinement but also indolence and vulgarity.[39]

The Japanese were also treated as a people of contradictory traits, and a constant source of fascination in American atlases and geography textbooks. At their best, they were "honest, ingenious, courteous, cleanly, frugal, animated by a strong love of knowledge, endowed with a wonderful faculty of imitation, and possessing a sentiment of personal honor exceeding that of any nature." But while they had an "instinct of progress" not unlike the Anglo-Saxons, they could also be "fickle, inordinately vain, and, at least the lower classes, exceedingly corrupt." Continued contact with European ideas—political, social, and religious—coupled with strong political leadership would ensure Japan's future of intellectual and industrial advancement.[40] In this manner these atlases combined in apparently contradictory ways the arguments of racial determinism and human possibility. By focusing on the world as having both a natural and a political order, these atlases introduced the American reader—most likely white—to his innumerable and exotic "fellow men," but also assured him the highest place in the hierarchy of civilization. Both Cram and Rand McNally rendered this argument graphically with visual typologies of the world's races that placed Caucasians at the apex (figs. 2.4 and 2.5). Rand McNally described the world through six racial portraits arranged in ascending, quasi-evolutionary order: Orang-outang, Malayan, Ethiopian, Indian, Mongolian, and finally Caucasian. More complicated but equally arbitrary was Cram's 1883 typology, which made Caucasians a category that included Europeans as well as "Chinamen."[41]

This racial division was graphically—and geographically—rendered in Cram's pictorial layout of the world entitled "The Four Quarters of the Globe" (fig. 2.6). Here we see a world arranged according to continental divisions that coincide with cultural divisions. As Martin Lewis and Kären Wigen point out, the convention of recognizing continents as the primary division of the world is itself a historically constructed practice. Just as the

ORANG-OUTANG, AND THE FIVE PRINCIPAL RACES OF MAN.

A—ORANG-OUTANG.   B—MALAYAN.   C—ETHIOPIAN.
D—INDIAN.   E—MONGOLIAN.   F—CAUCASIAN.

Figure 2.4   Racial typology, from Rand McNally's *Pictorial Atlas of the World* (1898), 165

concept of "Europe" replaced "Christendom" in the early modern era, by the seventeenth century the idea of a "world island"—encompassing Eurasia and Africa—was replaced by this newer construct, the "Four Quarters of the Globe." By the eighteenth century, the Ural Mountains had become both a real and a symbolic barrier separating Europe from Asia. With these changes the idea of continents as meaningful divisions of the world became more commonly accepted.[42] Here, Cram's diagram assumes that the continents are both physically and culturally constituted, that natural and human features coincide neatly in space. This use of continents as the definitive geographical division was reinforced by the fact that in many of these atlases no subcontinental coverage was given to Africa, Asia, and South America. These areas continued to be mentioned and understood primarily as continents rather than as nations, colonies, or the like. The graphic also entrenched the continental division through four illustrations that identify the geographical, physical, racial, and zoological particularities of each "quarter of the globe." This imagines for us a simpler, more essential metageography where continents correlate to race and physiography, itself suggestive of a kind of environmental determinism. Four types of men were placed in their distinct natural environments: in the upper left quarter was

an Indian in native garb who represented both North and South America;
to his right was a representative European, dressed in modern Western
clothing; in the lower left quarter was the representative Asian, and to his
right the model African. This illustration emphasized the typical features
of each geographical region by recalling a past where geographical and

Figure 2.5   Racial typology, from Cram's *Unrivaled Family Atlas of the World* (1883), 88.
Courtesy of the Harvard Map Collection.

Figure 2.6   Order of the world, from Cram's *Unrivaled Family Atlas of the World* (1883), 92. Courtesy of the Harvard Map Collection.

racial divisions harmoniously coincided. The message was clear: in the past each of the racial groupings belonged to a distinct region with uncontested boundaries. In an effort to make coherent the vast and complicated world, each quarter of the globe was represented by a continent: America, Africa, and Eurasia were each separated from one another by water, Europe and Asia by a great desolate land mass. Through all these graphic and cartographic devices, the complexity of geography was flattened and reduced to an uncomplicated structure of opposites and contrasts.[43]

In the 1870s and 1880s Rand McNally and G. F. Cram organized the world according to levels of "progress" attained by different races, nations, and continents. The physical information on these maps was limited and highly generalized, dividing the world into torrid and temperate zones, which further buttressed the ideology of environmental and racial destiny. Though the scope of these atlases was ostensibly international, most of the maps were devoted to the United States, reflecting a parochialism that would remain until a victory over Spain made unavoidable the question of growth overseas. Until then, these atlases brought the world home to Americans largely as spectacle, a distant reality that conformed to existing notions of racial and cultural hierarchy.

THE AMERICAN DEBUT IN A "SPLENDID LITTLE WAR"

The sinking of the Maine was a boon to Hearst and Pulitzer but also to Cram and Rand McNally. The Spanish-American War established a powerful relationship between war and maps that would grow exponentially during the First and Second World Wars. In each of these conflicts Rand McNally and Cram struggled to meet public demand for this new interest. The war atlas included different and more focused information than its peacetime counterpart, such as the status of the conflict, historical background, the belligerents' positions, and military campaigns. More generally, the companies constructed war atlases around a simple approach— victory or defeat—whereas world atlases were designed under more diffuse frameworks. War allowed bold articulations of the national interest and candid evaluations of other cultures and nations, while peacetime atlases generally assumed a more guarded approach. The war atlases ardently supported expansionism and immediately incorporated the drive for foreign markets and overseas territories into the cartographic narrative. Yet, in other ways war and world atlases were remarkably similar. Firms frequently relied on existing maps to illustrate the theaters of war, enhanced only slightly by the identification of battlefronts or fleet positions. Rarely were these war maps drawn on as small a scale as the Civil War maps printed in newspapers. Instead, the leading map companies repackaged standard maps and information in order to appeal to particular consumer desires.[44]

The motivation of appealing to consumer demand is doubly important if we are to understand the atlases of the Spanish-American War.[45] Comparison of these with the world atlases that preceded them suggests that in 1898 the nation "discovered" the world. In other words, these atlases abruptly begin to highlight the concern for foreign markets. Yet, as Walter LaFeber and others have argued, the events of 1898 were a culmination of rather than a break with the past. Why then do the atlases suggest otherwise? To answer this we must recognize that American cartography in the industrial era was in many respects a reactive medium. As David Woodward has argued, nineteenth-century firms such as Mitchell and Colton, which utilized copper engraving techniques, shunned the adoption of wax engraving and its attendant mass market in order to continue producing for a smaller, more exclusive audience.[46] Thus, the mass-produced, inexpensive war atlas was a novelty in 1898: Rand McNally and Cram were breaking ground by producing and packaging material primarily to capitalize on the frenzied interest brought by the news. Only secondarily were the cheap war atlases designed to address the cartographic contours of the war itself, a goal that would have demanded new maps drawn specifically for the geographical situation of the conflict. As a result the enthusiasm of these war atlases for foreign markets and expansionism generally was somewhat exaggerated. And though 1898 may not have been an all-important break for American foreign policy, its impact over geographic images available to the public was significant, its context and timing crucial.

The reconceptualization of the world around resources and commerce dominated the atlases as a motive for acquiring territory and more generally as an evaluative framework. The reorientation of foreign policy at the turn of the century accelerated this shift from a world of racial hierarchy where the United States stood apart to an economic world in which the nation was actively involved. This unconditional support for American economic and territorial expansionism is reflected in new maps and descriptions of Cuba, the Philippines, and other areas under American jurisdiction. Maps of these acquisitions were prominently and proudly featured in war atlases, indicating that the goal was not just to chronicle the conflict but also to announce and defend the territorial spoils of war. In Rand McNally's 1898 *War Atlas*, for instance, "vital information" printed about the Philippines, Cuba, and Puerto Rico focused on exports and resources rather than culture or race.[47] In the same year, Rand McNally introduced and evaluated the new territories according to their actual and potential commercial value. Typical of these profiles were histories of the Philippines that emphasized the islands' economic wealth but made only cursory mention of social or political life. Commercial profiles were also used to introduce America's emerging relationships to Hawaii, Cuba, and Puerto Rico, as well as the newly discovered riches of Alaska.[48] War atlases, like the contemporary issues of the *National Geographic*, were graphic arguments for the American mission abroad. Both media visually introduced the public to the new possessions and celebrated their potential contribution to American wealth. In both cases, the "science" of geography had translated controversial events and policies into matters of fact for middlebrow consumption.

This emphasis on natural resources was underlined in Rand McNally's 1899 *Atlas of Two Wars*, which narrated the military conflicts in the Philippines and South Africa. In analyzing the value of the Philippines to the United States, Rand McNally emphasized the untapped natural resources, including soil, hemp, sugar, tobacco, corn, rice, indigo, cacao, and cotton. As the authors put it: "As the islands occupy a commanding position as a base for commercial operations in the East and also possess vast stores of undeveloped wealth, with the end of Spanish misrule and the consequent removal of the disadvantages under which they have been struggling, their commercial advancement should be rapid and important."[49] Cuba was also introduced to Americans as a source of wealth, characterized in Rand McNally's *History of the Spanish-American War* as a vastly underutilized reserve that had been kept from Americans by corrupt Spanish tariffs on American imports. In one atlas the authors exalted the democratic nature of the American economic presence in Cuba by juxtaposing it with the cruel and rapacious leadership of the Spanish. The United States was an empire of positive principles that encouraged proper use and administration of the land, while Spanish greed hindered Cuban manufacturing. In general, these atlases described Cuba as unformed clay, and enthused that with "a wise and secure government" the island's growth and trade potential with the United States was virtually limitless. With protectorate control over Cuba the United States would no longer be subjected to unfair tariffs and instead would possess "The Key to the Gulf of Mexico."[50]

But the most suggestive of this political use of geographic and cartographic knowledge were not the descriptions of the Philippines or Cuba but those of Alaska. That the territory was even included in many of these "war atlases" reflects tacit approval of its inclusion in a larger expansionist movement. After all, by the turn of the century it had been part of the United States for more than thirty years. Though purchased in 1867, Alaska was only considered a shrewd addition to the nation after late-century discoveries of salmon, gold, copper, and forests transformed it from "Seward's Folly" into a rallying cry for further expansion. Rand McNally's *History of the Spanish-American War* described the limitless resources in Alaska and included a separate map of the territory. That Alaska was included in the atlas of "Recently Acquired United States Territory" even though its acquisition had been virtually ignored for three decades suggested a reconsideration of the territory as part of a larger potential American empire. This belated recognition was mirrored in contemporary geography textbooks, where the territory was absent from many maps of the United States before the 1890s but thereafter was prominently featured in both maps and text.[51] The Cram Company celebrated Alaska's wealth with an 1897 guidebook to its resources that would have been inconceivable a decade earlier. A foldout map highlighted the resource wealth of the territory, but even more telling was the description of its position relative to the nation's future.

> In acquiring the Alaskan territory . . . the United States moved its
> center, figured in geographical miles, not in area or population, as
> far west as San Francisco. The country now extends from about
> the 65th degree of longitude up at the far east corner of Maine to
> the 122d degree up at the far northwest tip of the Alaskan main-
> land. This is taking no account of the little island of Attu, 1,000
> miles out in the Pacific, beyond the Hawaiian group, which, since
> the purchase of Alaska, has really been our western land limit.
> The United States, therefore, may almost say with England that
> the sun never sets on its possessions.[52]

The attention paid to the physical dimensions of the American empire suggests an interest in identifying the actual territory under control, much like the then common maps that identified in pink or red the far-flung colonies of the British Crown. This interest was directly addressed by a new map that chronicled the territorial growth of the United States, classifying the acquisitions of 1898 as part of a long tradition that included the Louisiana Purchase, the annexation of Texas, and the Mexican cession (fig. 2.7).[53] Immensely popular—reproduced by Rand McNally, Cram, and Hammond—this map introduced the readers to the argument of both the war atlases and the war itself.

Through the science of cartography—which gave these changes a kind of authority and permanence—the divisive war had been transformed from controversial politics into immutable history. The special appeal of this expansionist map was reflected in a letter sent to the National Geographic Society years later.

> I should think it an excellent thing if someone would get out MAPS
> of all the EMPIRES of HISTORY, showing their GROWTH, something
> like maps of the U.S.—show the original COLONIES, the LOUISIANA

Figure 2.7 Rather than convincing readers to support the Spanish-American War, this map of the United States at the turn of the century (1900) simply framed the territories as the latest installments in a progressive, unfolding national history. Courtesy of the Harvard Map Collection.

PURCHASE, the GADSDEN PURCHASE, land acquired from MEXICO, SPAIN, and RUSSIA. In fact, show these empires at their beginning and at their greatest EXTENT, along with definite details of their development. Then it would be possible to say which of the historic conquerors brought the GREATEST AREA under his control.[54]

The writer celebrates the nation's status as an imperial power, but also reflects a deeper concern with the use of maps to make sense of and control space. In effect, the spirit of this map differs little from contemporary cartoons of the yellow press drawn to celebrate the nation's new, widely dispersed holdings. Many, perhaps faced with the enormous geographic and cultural distance between the Americans and their "little brown brothers," simply repositioned the Philippines off the American coast in order to legitimate control of the latter over the former (fig. 2.8). These kinds of drawings, like the map of American expansion, reflect a larger theme in the meaning of American popular cartography. Geared primarily to identifying and naming places, it is less interested in the nature of geographic relationships, distance, or topography. It should be unsurprising, then, that by

EUROPE: "MY GOODNESS, HOW HE IS MUTILATING THAT BEAUTIFUL MAP!"

Figure 2.8   One of many common illustrations "imagining geography" for the public during the Spanish-American War. From Charles Nelan, *Cartoons of Our War with Spain*, 2d ed. (New York: Frederick A. Stokes, 1898).

the end of the century the index to the world atlas—the primary means of translating an incalculably large reality into a finite realm—took up as much space as the maps themselves.

The introduction of wax into commercial cartography transformed maps from a craft into a commodity. This mass production technology facilitated access to these maps but also created a somewhat inflexible sense of what a map *should* look like that endured into the twentieth century. In the atlases themselves we find a consistent tendency to elevate the centrality of western Europe and the United States. While this alone is unremarkable, the way it occurs is significant for highlighting what are often entrenched assumptions about geography and space. By pointing to the way these practices were historically constructed—especially through the use of categories such as race, nation, and continent—we understand their power to draw for American consumers a world most would never see firsthand. In turn, this points to the subtle yet real relationships of science to nationalism and culture. No wonder maps have been described as "the silent arbiter[s] of power."[55] They legitimated what was controversial, made scientific what was historical, and naturalized what was human. By extension, world atlases quietly created a metageography for Americans by showing them a normative view of both the world and the map.

Science, Culture, and Expansionism in the Making of the *National Geographic* 1888–1900

Five days before Christmas 1905, the National Geographic Society held an elaborate dinner in Washington, D.C. With more than two hundred guests, among them explorers, diplomats, and members of Congress, the Society celebrated a membership that had grown from 3,400 to 11,000 in that year alone. Society President Willis Moore attributed this success to the character of its members, "the thinking, intellectual people of this city, of the nation, and somewhat from all nations—those who wish to keep abreast with the thought and activities of the world at large." For Moore, public interest in geography—as defined by the Society—fed the intellectual wealth of American civilization, which since 1898 had broken out of its continental confines and "leaped forward from this island to that, [until] today we find the free institutions of this country planted at the very doors of the Orient." In turn, the nation's new role abroad intensified the need for the Society to translate the meaning of these changes. The evening's guest of honor, Secretary of War William Howard Taft, affirmed Moore's convictions by exalting the nation's territorial and commercial gains, which sought "only to produce peace and prosperity the world around."[1]

By 1905 the *Geographic* was on its way to becoming one of the most ubiquitous sources of information and images about the world in American culture. The April 1905 issue—profiling the culture, politics, and resources of the Philippine Islands—was so popular that it had to be returned to press to meet the demand. But the monthly that became so enormously popular and influential in the twentieth century bore few traces of its original form. Founded in 1888, the *Geographic* was created to serve the community of scientists associated with the geographical work of the federal government, a category that included surveyors, topographers, statisticians, hydrographers, geologists, and explorers. The transition of their formal, narrow journal of scientific research into a heavily illustrated magazine of mass appeal has been explained through the arrival of savvy editors at the turn of the century, yet the magazine's transformation involved far more than just charismatic leadership.

What we find in the early *Geographic* are men working for the government who directly witnessed and aided the nation's political and economic expansion. The war between the United States and Spain gave the *Geographic* the exhilarating opportunity to cover international events and defend the nation's goals abroad, while bringing the distant and potentially enriching reaches of the new American orbit home to its readers. By taking advantage of this opportunity the magazine effectively enlarged the scope of geography to include not just physical science but also more ambiguous problems of politics and commerce that related to the nation's new international posture. The extensive use of photography, so central to the magazine's twentieth-century success, also began during the war. Thus, 1898 brought to the Society — as it had to mapmakers — a chance to translate the government's work abroad into material for popular consumption at home.

Considered together, these developments make the late nineteenth century a turning point for the modern form of the *Geographic*, and a closer look at this period reveals a dynamic relationship between science, culture, and American expansionism. During and after the war the community of scientists and government workers at the Society seamlessly joined their professional identities — whether as surveyors, hydrographers, statisticians, or diplomats — to the national interest. There is no evidence that the conflicts at the Society in these years reflected dissent over the nation's controversial mission abroad, and in fact the *Geographic* immediately became a place where the federal government's goals could be safely articulated. In this respect the Society at the turn of the century acted simultaneously as an organ of science and one of politics, which suggests that the scientific enterprise of geography was itself bound up with national concerns.

ORIGINS

The National Geographic Society was born in the intellectual ferment that swept Washington after the Civil War. It was one of a number of intellectual and scientific societies that served the community of scientists working for the federal government before the rise of universities.[2] Its founder and first president, Gardiner Greene Hubbard, was a wealthy attorney and philanthropist who had sponsored a number of scientific endeavors, among them Alexander Graham Bell's invention of the telephone. Hubbard was firmly embedded in the postwar scientific culture of the capital, three times voted president of the joint commission of the scientific societies of Washington, later organized as the Washington Academy of Sciences. As a result, he had no trouble recruiting other local scientists who shared his desire for a forum to organize and disseminate the extensive geographic research undertaken by the federal government. All five original founders and many early members would at some point work for the federal scientific bureaucracies.[3] With its clearly defined but limited constituency the Society grew modestly in the first ten years and granted only those local, "active" members a vote in the organization's business. Though the Society's bylaws encouraged a membership that was selective and influential rather than broad and national, President Hubbard hoped to attract nonspecialists like himself, particularly the political, military, and scientific leaders of the capital. In fact by 1895 the Society's lectures were drawing up to a thousand members, including students, armchair travelers, and other geographic enthusiasts.[4]

But in practice the *Geographic* remained a technical journal. W J McGee, an ethnologist who would serve as the Society's vice president and president, saw the magazine and Society as clearinghouses for geographic research and exploration, especially that conducted by the federal government. Speaking to the Board of Managers in 1896, McGee described the Society's lectures—many of which eventually appeared in the *Geographic* —as research and educational tools. He stressed the importance of maintaining geography's reputation as scientific rather than adventurous. In the lectures for the educated public, the speakers would be "actual explorers or original investigators who are known to treat geography as a branch of science." And in a remark that suggests something of the contemporary perception of geography, McGee reminded members that "excessive use of picture and anecdote is discouraged." A separate lecture series, designed only for practicing geographers, had an even more self-conscious agenda. In this forum McGee suggested that only a "recognized authority" be invited to address the Society's scientific members, adding "that superficial description and pictorial illustration shall be subordinate to the exposition of relations and principles."[5]

McGee argued to members that the purpose of the little journal was strictly to "convey new information and at the same time to reflect current opinion on geographic matters." The *Geographic* appeared erratically for the first few years, then settled into a monthly schedule of publication by 1896. The magazine was priced at twenty-five cents, while membership dues (which included Society privileges as well as a subscription to the *Geographic*) remained at five dollars per year until 1900, when they were reduced to two dollars. Upon opening the journal's terra-cotta brown cover a member found discussions of physical geography — studies of erosion, landform classification, weather patterns, coastal and land surveys, and exploration — illustrated by an occasional diagram, map, or photograph. Little attempt was made to gain new readers, as active members seemed content to maintain theirs as a local Society for working professionals. Thus although Hubbard had hoped for a society open to amateurs as well as professionals, in fact the *Geographic* was geared toward the latter. When Hubbard died in December 1897, membership was hovering at about one thousand, and the organization itself was two thousand dollars in debt. Into this breach stepped Alexander Graham Bell, who was initially reluctant to involve himself in the Society but felt obligated to uphold the legacy of Hubbard, his father-in-law.[6]

Nearly every account of the *Geographic*'s history tells the story of a dull, failing scientific journal transformed by the vision of Bell and the energy of young Gilbert Grosvenor, who was hired as assistant editor in spring 1899 and promoted to general editor in 1903. According to these interpretations, Grosvenor's leadership was the force behind the *Geographic*'s maturation to its modern, familiar form — one full of illustrations and appealing to the educated general reader — and its evolution into one of the largest and most influential magazines of the twentieth century. As Grosvenor himself wrote:

> The magazine's evolution from obscurity to phenomenal prominence makes a fascinating and often dramatic story. I hadn't been employed a year before I became involved in a fight for control of the magazine. A determined minority group wanted to publish it in New York City, sell it on newsstands, and omit all references to the Society. If these proposals had been accepted, the magazine inevitably would have become a commercial venture. But the *National Geographic* remains today the official journal of a nonprofit educational and scientific organization. . . . Once the fight for control had been won, I faced the task of evolving a magazine unlike any other in the world. It required an entirely new approach to the subject matter of geography.

There is no question that Grosvenor's leadership was daring and consequential — he survived an attempt within the organization to have him removed and the magazine published elsewhere — but the transformation of the *Geographic* involved more than any one individual's influence.[7] In his

brief account of the Society's early decades Philip Pauly pays special attention to this fight for control and stresses the roles of Bell and Grosvenor in
the magazine's development. In an increasingly specialized scientific culture, Pauly argues, the Society's popularity grew after 1900 because it ran
contrary to the trend toward professionalization, appealing instead to an
amateur tradition that was under assault in the reorganized American universities. Bell and Grosvenor aimed the Society at nonspecialists, and
broadened geography to include natural history precisely when that subject was being appropriated by other disciplines such as geology and anthropology.[8] While geographers struggled in the early twentieth century to
secure and defend even a small niche in the new American universities,
the National Geographic Society and its magazine managed to grow by
leaps and bounds, defining geography not as a subfield of geology but rather
as "the world and all that is in it." In fact the Society's popularity at the
turn of the century is something of a paradox. Many geographers—concerned about creating and defining their discipline within the university—
were consciously adopting an academic approach to the subject. By contrast, the *Geographic* adopted a general interest approach, and was wildly
successful. The further it moved away from its origins as a predominantly
local and scientific organization, the more attractive the Society became
to the general public.

Pauly is right to stress the academic context of the Society's early
growth, since the university science of geography was on shaky ground in
these years. Many blamed the field's uncertain status on its "popularization,"
or more generally on the tendency to associate geography with adventure
and exploration rather than scholarly investigation. William Morris Davis, one of the first American geographers to be given an academic post,
vocally criticized the geographical and exploration societies of the nineteenth century for diminishing geography's academic reputation. Davis believed that popular groups such as the Geographical Society of Philadelphia
and the Appalachian Mountain Club only limited geography's scientific
potential. Therefore Davis declined a place on the editorial committee of
the *Geographic* in 1901, arguing that the magazine had become excessively
popular and had lost its scholarly vision. Three years later Davis founded
the Association of American Geographers (AAG), created exclusively for
professional geographers and self-consciously styled against the National
Geographic Society. By January 1905 he made the separation complete by
resigning from the Society's Board of Managers.[9]

Much has been made of Davis's actions, perhaps too much. In resigning
Davis faulted the Society for geography's institutional weakness, but the
troubles facing university geography could not be attributed to a magazine

or to curious Americans who gathered in local or national geographical societies to discuss travel and exploration. This apparent "schism" between
professional and amateur geographers has been used to explain the success
of the Society as well as the unusual development of American geography. This portrait has also been widely accepted because both Davis and
Grosvenor—the leading figures on each side of the supposed divide—have
told essentially the same story through their words and actions. As Grosvenor recalled:

> the so-called professional geographers . . . to them the Geographic—
> it prints too much about flowers and animals and so on. They say
> that isn't geography. They're very foolish. . . . When I started in, I
> said we must interest the schools. We must print things they'd be
> interested in. We got a man named William M. Davis who was the
> leading geographer of the period to come on the Board. Davis was
> a very fascinating talker, and he got a strangle hold on the geogra-
> phers of the country. Which I didn't learn until afterwards. Well,
> he didn't like the way I was running the Geographic, and after be-
> ing on the Board for about two years, he retired and started an-
> other Society which is called now the Association of American
> Geographers.[10]

The division between amateur and professional is also an explanation that
fits easily with the historiography of social science. Thomas Haskell, investigating the late-century transition from amateur to professional social
science, identifies this as a period when "the professional defines himself
largely by the practices he rejects."[11] And yet while this schism existed—
some geographers were anxious to dissociate themselves from the National
Geographic Society, as we will see in chapter 4—it does not explain why
so many of the original members of the AAG continued their association
with the Society. Davis himself continued to give lectures at the Society
and to solicit funds on the organization's behalf.[12] The schism between professors and amateurs at the Society also does not account for the fact that
in 1905 seven of the sixteen editors of the Geographic were also members of
its purported rival, the university-based AAG. The magazine's transition at
the turn of the century involved more than the conflict between amateur
and professional identities.[13]

THE CRUCIBLE OF WAR

Far more consequential for the Geographic was the tacit assumption—by
professionals and amateurs alike—that geographical knowledge was linked

to the health of the nation itself, a precondition for vigorous nationalism. Early in January 1896, newly appointed editor John Hyde—a statistician for the Department of Agriculture—celebrated the close relationship between the state and the Society. Introducing readers to the new monthly series of the *Geographic*, Hyde promised it would embody the national experience by reporting not just on the work of the federal scientific bureaucracy but also on areas of the Western Hemisphere in which Americans had "an exceedingly keen and friendly interest." For Hyde, this particular role would make the new *Geographic* "American rather than cosmopolitan, and in an especial degree . . . National." Though he was certainly not the force that Grosvenor would be at the magazine, Hyde believed—like his successor—that the *Geographic* was responsible to the public and the national interest. In defending its new mission, Hyde claimed that "to possess a knowledge of the conditions and possibilities of one's own country is surely no small part of an enlightened patriotism, and to the patriotic impulses of the American people no appeal was ever made in vain."[14] Two years later, the events of 1898 allowed the Society to put into practice what was already a principle at the magazine, that geographical knowledge was a tool of nationalism. The Spanish-American War encouraged the men at the magazine to broaden geography to include newsworthy and controversial problems such as race, commerce, and colonialism. As Society president Moore had put it, by 1906 the scope of the Society had become "the thought and activities of the world at large." This willingness to cover the war as a meeting ground of national and scientific interest was further fueled by the contemporary intellectual context of academic geography. As argued in chapter 4, in order to be relevant and useful to both the natural and the human sciences, geographers widened their charge to include not just the physical landscape but also assessments of human progress in that landscape.

Among those at the *Geographic*, many of whom had long worked for the government, this framework was quickly accepted, as manifested in the magazine's dramatic yet unopposed transition from a journal of physical science to one that extended to include political and economic concerns. W J McGee, who would later become an editor at the *Geographic*, had in 1896 defined geography as "the causes and conditions by which human progress is shaped."[15] To McGee, this discovery of causal principles widened geography from an inventory of the physical landscape to a dynamic science of human response to the environment. As he explained,

> the geography of the future will be devoted primarily to research
> concerning the forces of the earth, including those affecting

peoples and institutions as well as those shaping land-forms and molding faunas and floras, and that industries, arts, commerce, laws, governments, religions, even civilization itself, will eventually fall within the domain of geography. The prediction is easy and safe because the geography of the present is already on the higher plane with respect to the inorganic part of its object matter, is well advanced toward this plane with respect to the evolution of organisms, and looks up to the same plane with respect to the courses and causes of human organization; the fulfillment of the prediction will be simply the consummation of present progress.[16]

Pejoratively termed "environmental determinism" by later generations, this explanatory framework was eagerly endorsed by the Society at the end of the nineteenth century. Consider the theme chosen by the Society for its 1897–1898 lecture series: "The Effect of Geographic Environment on the Civilization and Progress of Our Own Country." For President Hubbard, this theme explained why progress had come to certain areas but left others unimproved. Until the discovery of the New World, he wrote, "natural causes" had been the primary influence over human civilization. Thereafter, more advanced civilizations had come to subdue their environment and thereby to liberate themselves from this influence. The Spanish-American War encouraged those at the *Geographic* to think about historical development in light of this relationship between natural environment and human behavior. This formulation enabled the *Geographic* to valorize the nation's work abroad, which would aid those less able to master their own environment.[17]

This redefinition of geography is reflected in the magazine's changing content. Before 1898 the *Geographic* rarely addressed issues not directly related to the physical landscape.[18] But by the time President McKinley asked Congress to declare war against Spain in April of that year, the magazine had immersed itself in this political conflict by devoting entire issues to Cuba and the Philippines. Between 1898 and 1905, the *Geographic* printed ten articles related to the situation in Cuba, six accompanied by a map. This is particularly striking given that the island had never been mentioned in the magazine before the war, despite longstanding problems between Spain and Cuba and the proximity of the latter to the United States. Even more striking is the attention paid in the magazine to the Philippines, though these islands too had never been discussed before 1898. Thirty articles, four of them including a map, were devoted to the Philippines between the outbreak of the war and 1905. The increase in attention extended to Puerto

Rico (the subject of twelve articles from 1898 to 1905) and the tiny island of Samoa (covered four times from 1898 to 1900) as well. In fact, the *Geographic* paid more attention to these islands at the turn of the century than to any other part of the world, with the exception of Alaska and the Arctic, both regions that appealed to public interest in exploration. Admittedly, the romance of exploration captivated many *Geographic* readers after the turn of the century, but this tradition has obscured the degree to which the magazine vocally participated in the debate about American expansion overseas.

The impact of the Spanish-American War on the magazine was further heightened by the *Geographic's* historically close relationship with the federal bureaucracies. In 1899 the magazine's board of fourteen included only three who worked outside the government; the remaining eleven were drawn from the USGS, the Hydrographic Office, the State Department, the Department of Agriculture, the U.S. Treasury, the Bureau of Ethnology, and the leadership of the armed forces. These men were also among the most frequent contributors to the *Geographic,* and they welcomed the opportunity to publish commentaries by illustrious political officials such as Secretaries of War Elihu Root and William Howard Taft, Secretary of State John Watson Foster, Assistant Secretary of State David Hill, and a number of diplomats. And because the *Geographic* could not yet support correspondents, it relied on federal agencies for maps, census figures, surveys, and other geographic information. O. P. Austin, a frequent contributor to the *Geographic* and Chief of the Bureau of Statistics in 1898, remarked on the rush of inquiries the government received during and after the war for the latest information on America's new territories. Here the close relationships between the magazine and the government clearly worked to the former's advantage. For instance, A. W. Greely had fought in the Civil War and explored the North Pole before heading the Army Signal Corps at the turn of the century. So when the *Geographic* needed a map of Cuba, Greely simply asked the Corps to supply one. A few years later the Society developed its own map of the Philippines, modeled — as many others would be — on drawings from the War Department and the Army Corps of Engineers.[19]

GEOGRAPHY IN SERVICE TO THE STATE

Writers for the *Geographic* were also influenced by their work for the state as military officers, surveyors, ethnologists, statisticians, and in other capacities. As Chief Signal Officer of the Army between 1898 and 1902, Greely brought the far-flung colonies closer to America by overseeing the installation of extensive telegraph lines in Cuba, the Philippines, Puerto Rico, Alaska, and China. After the United States took administrative control of the Philippines, Taft and other members of McKinley's Philippine Commission were sending regular reports to the Society for publication in the *Geographic*. Similarly, the extensive census taken in Puerto Rico, Cuba, and the Philippines involved a number of representatives from government agencies or the military who were also members of the National Geographic Society and writers for the magazine.

In general, the men behind the *Geographic* were eager to offer their skills in service to the state. Given their backgrounds it is not surprising that they saw the goals of the nation as coinciding with those of their own areas of expertise. In their view, a strong national presence internationally would strengthen their fields, just as a firmly grounded science might enhance the nation's position abroad. As state servants, they easily redefined geography as consistent with commercial and political expansion, just as before the war the government's work in geography had been defined through different state goals such as topographic surveys of the West or meteorological studies. In 1898 the Society was "national" not so much in scope or membership as in identity. While the magazine could have systematically opposed the acquisition of territories, this was unlikely given the close working ties between these men and the federal bureaucracy. At times the Society and the *Geographic* functioned very much like an arm of the government.

This is most clearly manifest in the *Geographic*'s explicit and consistent defense of the American mission at century's end. The tone was set by the *Geographic*'s "Cuba" issue of May 1898, introduced with a full-page portrait of Captain Charles Sigsbee, the commander of the battleship *Maine* that had exploded in Havana harbor and sparked the American sentiment for war. In glowing prose the magazine cited not just Sigsbee's military bravery but also his contributions to hydrography. The Society held him up as a man who embodied its ideals of objectivity and national honor, as disinterested in his pursuit of science as he was committed to protecting the nation's role abroad. In this regard Sigsbee was just the man the new National Geographic Society could, and did, honor with an elaborate ceremony in the nation's capital.[20]

Over the next several years the magazine continued to champion the government's work abroad, specifically in the Philippines, Cuba, and Puerto Rico, and the Anti-Imperialist League would find little sympathy in the pages of the *Geographic*. The sole exception to this rule was Henry Gannett's brief caution in December 1897 against the "Annexation Fever" that was sweeping through the "civilized world." Gannett advised the United

States to resist this impulse, for any addition "to its numbers will reduce the average civilization, and consequently the strength and industrial capacity of its people." Gannett lamented the costs brought by Alaska and warned that Hawaiian annexation might bring similar disappointments. But this article stands out as the exception, a position Gannett himself repudiated in subsequent months by his celebration of American commitments in Asia and the Caribbean.[21]

The *Geographic's* defense of American expansionism invoked expanding trade and native uplift, both of which were fortified by the accusation that Spain had failed to administer its possessions properly. Spanish incompetence fueled "liberation movements" in both Cuba and the Philippines that echoed America's own revolutionary tradition in their resistance to oppressive taxation and tyrannical rule. Given this characterization, the time was ripe for American intervention, and at no time after the war began did the *Geographic* voice anything but unqualified support for these goals. In May 1898, Robert Hill of the United States Geological Survey introduced Americans to the situation in Cuba by comprehensively outlining not just its physical dimensions—which would have been typical given his position and the magazine's scope—but also the island's history, people, commerce, and customs. Never losing an opportunity to chide Spain for its shortsighted and cruel colonial policies, Hill held out the hope of American rule in Cuba:

> In all history no other country has presented such an unfortunate exhibition of misgovernment. Perhaps ere this article reaches the reader the great government which stands for the highest type of humanity and whose every interest—commercial, hygienic, and strategic—calls for a cessation of Spanish misrule, will have made its influence felt and established a permanent peace upon the island.[22]

In the next issue, the *Geographic* introduced readers to the Philippines through the eyes of Colonel F. F. Hilder, a soldier, geographer, and ethnologist well suited to the work of the Society. Born in Britain in 1836, Hilder served in the British army in India and Africa until an injury ended his military career. He then emigrated to the United States to seek his fortune in Latin American markets as a representative for Remington Arms. This business interest—which took him on extended visits to the Philippines—together with his ethnological research gave him the authority wanted by the Society for its first profile of the islands.[23] Hilder began by describing the agriculture, topography, and climate of the Philippines as well as its products, city life, and history, all of which reflects the expanding scope of the magazine. Echoing the sentiment of writers for the Cuba issue, Hilder sympathized with Filipino discontent under Spanish rule by pointing to the "more civilized natives and particularly the half-breeds, who are sufficiently educated to crave for greater freedom."[24]

The *Geographic* did not simply frame American commitments as an act of altruism. Nearly every article related to the territories cited the resources and markets that would follow, in many instances as a way to persuade the ambivalent reader. Hilder followed up his survey of the Philippines with a story on gold mining on the islands two years later. John Hyde, the editor of the *Geographic*, pressed American intervention in Cuba in the May 1898 issue by rhetorically asking whether it was "any wonder that, entirely aside from the humanitarian considerations . . . some justification for such intervention should have been found in the well-nigh total paralysis of our commercial relations with that once extensive and profitable market?"[25] Hyde was equally convinced of America's proper role in the Philippines, and justified a transfer of control of the islands as a way to redress the trade imbalance with the United States. His position was amplified by its placement alongside an editorial reprinted from the *Financial Review* that almost forcibly endorsed the nation's need to control new colonies.

> What claim can any power advance, or by what right can they
> demand that our government evacuate these islands? None! . . .
> [T]his war will result in untold advantages to the United States.
> Our aim was to banish Spain from the Western continent and free
> an oppressed people. Our reward is the unexpected acquirement of
> territory and control of the trade of the Antilles, and a foothold in
> the development of the Orient.[26]

Dean Worcester, a scientist who would later serve on the Philippine Commission and as Secretary of the Interior of the Philippines, framed similar sentiments with only slightly more detachment during the war:

> Should the Philippine islands become a permanent possession of
> the United States or of any other civilized nation, the problem of
> giving them good government and of developing their enormous
> latent resources will be by no means a simple one, although it will,
> in my judgment, be one that will richly repay successful solution.
> Spain has never seriously attempted to solve it. From the time of
> its discovery until now the archipelago has been one vast plunder-
> ing ground for her hungry officials. She has conquered so far as
> greed of gain made conquest desirable or safety demanded it, but
> there she has stopped.[27]

As another writer put it, "the total returns are nothing like what they should be" among these "fertile . . . and most neglected colonies." From the Bureau of Statistics, Austin reported the "present consuming power is, in round terms, one hundred million dollars." (Taft reported in 1907 that annual exports to the territories reached approximately $66 million.)[28] Some even expressed the hope that the manufacturing wealth of the Philippines might quickly surpass that of Japan. The Society welcomed confirmation of this hope from a representative of the European imperials themselves, Major A. Falkner von Sonnenburg, of the Imperial German Army and the Former Military Attache at Manila. In the *Geographic* this German insider reassured Americans that the Spanish had been historically weak in the Philippines, but that a "stronger, more energetic, and more gifted race, with unlimited financial resources, may do in the future all that the former masters failed to do in three centuries."[29]

The *Geographic* prepared its readers well for American stewardship of these territories. But this expansion of responsibility came simultaneously with a recognition that they were inhabited by people who were utterly different and possibly less advanced than Americans. To allay these fears the *Geographic* repeatedly invoked the second goal of American internationalism: progressive uplift. John Barrett, the former American Minister to Siam, assured Americans that attention to the Philippines would not be wasted:

> Of the people who inhabit the Philippine Islands I can say, after extended acquaintance with them, that their good qualities far outweigh their bad qualities. When they are not misled or misguided by ambitious leaders in regard to America and the American people, they will become peaceful subjects of our government. When once order is fully established, there will be little or no spirit of insurrection manifesting itself, except where now and then, as in any land, some headstrong, unscrupulous leader may endeavor to resist the government. The majority of the Filipinos are far above the level of savages or barbarians and possess a considerable degree of civilization. It is the small minority that are wild and untamed in life, habits, and system of government.[30]

PROGRESS THROUGH PHOTOGRAPHY

The descriptions of the natives of these new territories were made even more persuasive through the use of photography. Though photographs had appeared in the magazine before the war, it was the Cuba and Philippine numbers that initiated the heavy use of illustration, a hallmark of the twentieth-century *Geographic*. In the Alaskan Klondike issue of April 1898, for instance, we find a few photographs of explorers alongside shots of the landscape and wildlife of the region. Just one month later, the Cuba issue opened with a portrait not of a geographer or an officer of the Society but of a military hero. The featured article on Cuba that followed included routine topographic maps and diagrams interspersed with village and street scenes. Hilder's feature on the Philippines in June 1898—originally a special lecture given to the Society in May—was also heavily illustrated. Though the photographs were published courtesy of *Leslie's Weekly*, before long the Society would have the resources to hire its own staff photographers or purchase illustrations for its own use. Not only did the magazine print more photographs during the war, but it also began to include human subjects in addition to physical landscapes. Village scenes, cityscapes, portraits of farmers and more primitive natives filled the Philippine issue, greatly expanding the range of appropriate subject material in the *Geographic* (fig. 3.1).

This intensified and widened use of photography marks a critical shift for the magazine. Most commentators date the origin of the *Geographic*'s modern photographic tradition to 1905, when a Russian explorer offered fifty unsolicited photographs of Lhasa to the magazine in exchange for acknowledgment of his authorship. Grosvenor recalled insisting that the photographs appear with minimal text, a decision which—perhaps because it met with resistance from other editors—has been understood as a turning point. But while Grosvenor may have been responsible for introducing the photographic essay in 1905, photography had become central to the *Geographic* much earlier. In a letter to Grosvenor in March 1900, Bell stressed the readers' enthusiastic response to a photograph of partially nude African women and encouraged his son-in-law to seek out and print this type of image, both exotic and human-centered (fig. 3.2).[31]

Significantly, among the most heavily photographed subjects in the magazine at the turn of the century were the Philippines, featured in June 1898; May 1903; March, June, and July 1904; and April and August 1905. These photographs of the Philippines—like the narrative text—served to inform, to entertain, and to defend the nation's new interventionist posture. In this regard the photographs functioned much like contemporary exhibits of Filipinos found at the Pan-American and the Louisiana Purchase Expositions. In fact, the men recruited to organize the Philippine exhibits at both these events were both closely connected to the National Geographic Society. Hilder, former Secretary of the Society and contributor to the *Geographic*, was appointed to gather material from

Figure 3.1 This photograph, published in 1898 during the Spanish-American War, was among the first of women to appear in the *National Geographic*.

Figure 3.2 This image of two "Kafir matrons" of South Africa was the first full-page photograph of barebreasted women to appear in the *National Geographic* (1900). Their nudity is, significantly, offset by their role as mothers.

the islands for the United States government's exhibit at Buffalo's Pan-American Exposition in 1901. The organizers of the Buffalo Exposition charged Hilder with finding materials that would illustrate the resources, government, and mode of life of the Philippines as well as the character of the people themselves. The resulting exhibit, including an eleven-acre "Filipino Village" populated by a hundred natives, was one of the most popular attractions at the fair.[32] The National Geographic Society heartily endorsed the exhibit for its authentic recreation of "a typical Filipino village inhabited by genuine natives," and was especially pleased with the exhibit's ability to boost public opinion of the American role on the islands. This was particularly critical given that the United States was no longer waging a "splendid little war" against Spain but rather a bloody and protracted campaign against former allies in the war, Emilio Aguinaldo and the Philippine rebels.[33]

The public's enthusiastic reception of the Philippine display emboldened the organizers of the Louisiana Purchase Exposition three years later to mount a qualitatively larger exhibit of "the barbarous and semi-barbarous peoples of the world, as nearly as possible in their ordinary and native environments."[34] That they chose W J McGee to run the exhibit was significant, for he was firmly convinced that America's recent experience constituted the fulfillment of manifest destiny. An ethnologist who had worked under John Wesley Powell at the United States Geological Survey and the Bureau of American Ethnology, McGee was heavily involved with the National Geographic Society, first as an editor of the *Geographic* and then, at the time of his recruitment to St. Louis, as president of the Society.[35] At a cost of more than $1 million, McGee's Anthropological Exhibit included nearly one thousand Filipinos on a "Philippine Reservation," as well as other different nonwhite "types": African, Patagonian, Ainu, and Native American. The goals of this exhibit again closely paralleled those of the *Geographic*. Furthermore, the Reservation and the *Geographic* both addressed the commercial potential of the region as well as the moral responsibility brought by annexation. What Robert Rydell has said of the Exposition — that it "gave a utopian dimension to American imperialism" — also describes the *Geographic*'s depiction of the Philippines in the aftermath of the Spanish-American War. The Society applauded McGee's Philippine Reservation as an authentic reproduction of life in the islands that would not just entertain the public but also facilitate mutual understanding between Americans and Filipinos at a critical moment in the relationship.[36]

In a related way, the *Geographic* consistently highlighted the progressive nature of the American presence in the Philippines. This is starkly demonstrated by Grosvenor's seminal April 1905 feature, "A Revelation of the Filipinos." Predictably, the issue included photographs of the "native types" that closely paralleled the style and content of photographs from the Philippine Exposition in St. Louis. Yet the magazine opened not with these but with portraits of "civilized" Filipinos who had adopted the dress, posture, and character of Western culture. Rydell describes the Philippine Exposition in St. Louis as designed to highlight "the attributes of civilization—social and political order, education, and commerce—that the federal government considered essential to the future well-being of the islands." Similarly, the *Geographic* article presented Americans with an impressive array of achievements in the Philippines that balanced even the exotic and stark portraits of natives (fig. 3.3).[37]

Photography was also used to document the progressive impulse in Cuba, which in turned helped justify the nation's presence and commercial gain. The power of American science was brought home vividly to readers through the juxtaposition of images of life in Cuba before and after the improvements of the Army Corps of Engineers, making it difficult to deny the benefits of American occupation (figs. 3.4–3.7). The photographs described Cuba and the Philippines as familiar—cleaner, more modern, abundant in resources, and commercially sophisticated—but at the same time situated the territories as sufficiently distant, both geographically and culturally, to be nonthreatening. In their survey of the recent history of the *Geographic*, Jane Collins and Catherine Lutz identify this tendency to present simultaneously the foreign and the familiar—the dangerous and the tame—as one of the persistent dynamics of the magazine during the cold war, and this trend can be identified as early as the turn of the century when the magazine first introduced a human and political dimension.[38] This sense of balance enabled the *Geographic* to tap the adventurous spirit of the reader but also to maintain a sense of control. Moreover, it is significant that the photographic tradition in the magazine began as a way to demonstrate the power of America's mission abroad.

This attention to improving the territories continued throughout the first ten years of the twentieth-century *Geographic*, through both text and photographs. The role of Henry Gannett, Chief Geographer of the USGS, treasurer of the Society and later its president, is emblematic here. Though Gannett had been wary of American expansion in 1897, his subsequent involvement in the Philippines and Cuba reveals his support for the nation's work. In fall 1899 he supervised a census in Cuba with the War Department, which the following spring was used in the *Geographic* to illustrate the progress made on the islands. The *Geographic* summarized the findings by indicating that administrative control of the island would remain safely "in the hands of the native white Cuban when the United States withdraws

Figure 3.3   This photograph of young boys in the Normal High School of Manila
accompanied William Howard Taft's highly popular article on the Philippines
(August 1905). Its caption, "The Right Road to Filipino Freedom," suggests the
Geographic's vision for the future of the islands. [Photograph credited to Underwood
and Underwood, New York.]

from the island," thereby preventing Cuba from becoming "a second Haiti."
A few years later Gannett supervised the Philippine Census, assuring *Geographic* readers that the war was over and the people pacified, that they had
"the utmost respect for Americans, a respect rapidly ripening into confidence and affection." With this, the potential gains from the islands "under
civilization and careful and intelligent cultivation are almost infinite."[39]

The *Geographic* also publicized the work of the Philippine Commission,
appointed by McKinley in 1899 to administer the islands. These articles

Figures 3.4 and 3.5 "Before and after the arrival of American Officers"—*National Geographic*'s visual demonstration of "American improvements" in Cuba's Colon Park (1902).

Figures 3.6 and 3.7   Two Cuban street scenes documenting "American progress in Habana." From *National Geographic* (1902).

continuously updated readers on the status of the territory in a particular way; like the photographs, the articles suggested a primitive and foreign quality but reassured the reader that this place eagerly awaited improvement. If the Philippines had been disease-ridden, they now demonstrated medical and sanitation improvements. If there were dozens of languages spoken on the island, this situation was being corrected with the welcome introduction of English-language instruction in villages and towns. If the Filipinos had failed to maximize agricultural possibilities, "fortunes await[ed] American market gardeners in the suburbs of Manila." The *Geographic* argued that the territories were modernizing, and used this as a rationale to champion to the public the benefits of scientific expertise, especially that developed by the surveyors, statisticians, engineers, and others who had themselves been historically active in the work of the National Geographic Society.[40] It was in fact the densely illustrated April 1905 feature on the Philippines—with hundreds of photographs taken from the War Department and the Philippine Commission—that fueled the largest proportional membership gain in the Society's history.

INTO THE AMERICAN CENTURY

The *Geographic*'s support for the war was not unique; contemporary magazines, newspapers, atlases, and books from 1898 and 1899 reflect the widespread enthusiasm for the nation's military adventures. But what makes the *Geographic* interesting is its ability to marry the imperatives of science with popular interest. Addressing the Society in spring 1899 on "National Growth and National Character," W J McGee outlined the history of the United States in a way that affirmed the need for further territorial growth, a task for which Americans were well fitted by virtue of racial and cultural superiority. After outlining the progressive history of the United States in terms of expansion, McGee boasted that "if Cuba and Puerto Rico, Alaska and Hawaii, and Luzon and her neighbors do not make America the foremost naval and shipping nation of the earth within a quarter-century, then experience stands for naught, history is a delusion, civilization a failure, and enlightenment a farce."[41] Just as weighty were the claims of John Barrett, the former minister to Siam, writing about the Philippine Islands and their people in the January 1900 *Geographic*:

> we became, by the battle of Manila Bay and the occupation of the
> Philippine Islands, the first power of the Pacific, for the control of
> which we seem to be destined by the great influences which shape
> the politics of the world and develop nations for mighty responsi-

bilities. If we bravely perform our duty in the Philippines, establish peace and order, give the people a large degree of autonomy, spread the influence of our free institutions and hold there a position of commercial and strategic advantage for the advancement and protection of our vast growing interests in the Pacific and far East, we shall be forever the first power of the Pacific and of all the world. If we are laggards now, we shall be laggards until doomsday. If the war and occupation of the islands costs us hundreds of millions of dollars now, another war, which would inevitably come in the future if we should try to regain the position lost by withdrawing from the islands and to lead in the merciless race of nations for material and moral supremacy, would cost us ten times as many million dollars.[42]

The experimentalism afforded the *Geographic* by the war and its aftermath encouraged the magazine's editors—particularly Bell, Grosvenor, and Hyde—to think about what the magazine might become. By mid-1899 editor John Hyde was optimistic about the Society's future:

It is doubtful if the study of any branch of human knowledge ever before received so sudden and powerful a stimulus as the events of the past year have given to geography. . . . There is not one of the new territorial possessions of the United States the geographic conditions and economic possibilities of which have not already been discussed, under the auspices of the Society, by distinguished men who are thoroughly familiar with them from personal observation and research, and it would be almost impossible to devise a means of more effectually promoting the Society's objects than by the delivery of these and other entertaining and instructive lectures in all the large centers of population.[43]

Hyde was pleased with the *Geographic*'s demonstrated ability to make available and comprehensible to the public the government's technical research. To Hyde, the Society and its magazine had become "an agency popular and yet authoritative," a role it began to institutionalize after the war. In 1899 the Society planned a popular lecture series that would include the Alaskan and Venezuelan boundary disputes, conflicts in the Transvaal and Manchuria, progress in the Philippines, Arctic exploration, and the prospect of a canal in Nicaragua. These new lectures were designed to be "attractive and interesting" and to avoid an overly "technical character." Lantern slides to illustrate the lectures, which McGee had discouraged as late as 1896, would now not only be tolerated but encouraged. Yet another course of lectures that had previously discussed American expansion would focus after the war on the growth of "the foremost nations of the world."[44]

By 1900 President Bell was capitalizing on the *Geographic*'s recently enlarged scope and focus by selling the magazine as preeminently one of current affairs. Grosvenor even hoped to place correspondents in the Philippines and China so the magazine might offer firsthand accounts of life and news in the nation's newest possessions. Both Bell and Grosvenor understood the significance of the events of 1898, and considered the coverage of the Philippines a model that would shape coverage of future topics, such as the outbreak of war in the Transvaal in 1899 or the Russo-Japanese War in 1905.[45] The magazine followed suit. Throughout the early twentieth century articles appeared on the Philippines and Cuba. Taft, named head of the Philippine Commission in 1899 and Secretary of War in 1904, vigorously defended the nation's role there against critics of colonialism. Also heavily covered were other new areas of political and economic influence for the United States: the Open Door in China, the boundary dispute and gold discoveries in Alaska and the Yukon, and the Panama Canal.[46] Looking back, it was the Spanish-American War that first suggested to the editors that geography included the realm of human—even political—interaction.

As we will see in chapter 7, the magazine's success in the twentieth century also owes much to the perseverance and imagination of Bell and Grosvenor. They were able to capitalize on human curiosity and channel it into an enterprise that quickly became an American institution. Their keen eye for creating and finding geography in interesting places—using exploration, zoology, aviation, anthropology, entomology, and meteorology—captivated millions. In this regard the *Geographic* is unrivaled, and owes much to the perseverance and imagination of these two leaders. Grosvenor's stature was also enhanced by the trials he endured with the editors after the war. While Grosvenor was in Europe on honeymoon, John Hyde and the magazine's executive committee tried to move the magazine to New York, and since then Hyde has been characterized as the symbol of the magazine's staid, technical, and narrow early history. But the conflict between these two men was more a difference of personality and style than of substance, and in fact Hyde's suspicion of Grosvenor had more to do with the latter's lack of experience in science and government than with any supposed rift between amateurs and professionals. The conflict between Grosvenor and the older members of the Board also had much to do with the relationship between the Society and the magazine. Some board members were eager to see the two separated, and the magazine sold on a sub-scription basis rather than one that tied readers to membership, while Grosvenor insisted that neither the magazine nor the Society could stand alone. Also a source of conflict between the two factions were stylistic ap-proaches such as the proper spelling of Puerto Rico (versus "Porto Rico") and Beijing (versus "Peking"), which some have made out to be symbolic of the division between amateurs and professionals. While there is some validity to this interpretation, it overlooks the shared internationalist sen-sibility among both groups and their common involvement in the nation's activity abroad.[47]

Consider the complete title of Grosvenor's famously popular feature of April 1905: "A Revelation of the Filipinos: The Surprising and Exceed-ingly Gratifying Condition of Their Education, Intelligence, and Ability Revealed by the First Census of the Philippine Islands, and the Unex-pected Magnitude of Their Resources and Possibility for Development." Like dozens that came before, this article framed the Philippines as ethno-graphically distinct from the United States, yet made familiar by both the Filipinos' willingness to accept American improvements and the substan-tial economic benefits to be gained. This pattern of taming the world and making it safe for American popular consumption would persist in the *National Geographic Magazine* for decades.

The *Geographic*'s maturation at the turn of the century reflected the na-tion's military, political, and commercial commitments, and in this regard geography grew up around national imperatives. The Spanish-American War helped the magazine showcase its work in a way that resonated deeply with the public, first because it involved human subjects and second be-cause it endorsed the nation's economic and political position abroad. The conflict also allowed the men at the *Geographic*—Bell, Taft, Hyde, Gros-venor, Gannett, and McGee—to unite science with national interest, making geography both a tool of expansion and a medium of middlebrow culture. To a great extent, then, the roots of the modern *Geographic* lie in the events of 1898.

4

Creating the Science of Geography
1880–1919

The staggering territorial growth of the United States in the nineteenth century solidified an enduring myth about expansion as the destiny of the nation. This in turn placed a premium on knowledge of the trans-Mississippi region, and antebellum American geography prospered on the basis of its ability to satisfy this demand. As chapter 3 suggests, through much of the century geographical knowledge was defined broadly and utilized by a wide community—including scholars, explorers, bureaucrats, politicians, and amateur explorers—housed either in the government or in organizations such as the National Geographic Society and the American Geographical Society. Chartered in 1854 by the New York legislature, the American Geographical Society's members were especially concerned with the progress of the nation's westward movement and focused their antebellum efforts on finding the best rail route to the Pacific. Many influential members also used the Society to propose expeditions or internal improvements in the West. George Bancroft, a founding member of the Society and its president from 1851 to 1854, was himself an ardent supporter of President Polk's expansionism and the Mexican War. Bancroft was a historian, and most of his fellow AGS members were not geographers but printers, editors, and writers. This diverse membership meant that officials and leaders in government, business, and education were exposed to ideas involved in the exploration, surveying, and mapping of the American West.[1] And while members of the AGS were certainly familiar with David Livingstone's travels in Africa, their attention was more commonly directed westward, following closely the work of the United States Geological Survey in the 1860s and 1870s.

In this regard the nineteenth-century American geographical vision can be characterized as more parochial than its British counterpart. Still, there are important parallels between American and British geography in the nineteenth century. The Royal Geographical Society, founded in 1830 by a group of explorers and travelers, served many interests: as a storehouse of knowledge, as a pressure group, and as a promoter of the study of international affairs. This broad agenda reflected its professional, if not ideological, diversity, a quality it shared with the American Geographical Society. But by the late nineteenth century it was clear in both the United States and Britain that geography was no longer simply a tool of exploration, data gathering, and mapping. In 1887 Halford Mackinder declared that all but the Poles had been explored, thereby heralding a new age where the imperative to expand clashed with the fact that most regions were already claimed. Together with Turner's musings on the closing of the frontier, Mackinder's declarations had particular resonance for geographers, especially because they coincided with the expansion and professionalization of the American university. The growth of university disciplines in the 1880s and 1890s, together with symbolic declarations of the exhaustion of spatial frontiers, demanded a reconceptualization of geography as a more analytic, scientific body of knowledge. Institutionally and intellectually, this was a difficult reorientation for geography. Its formal practitioners openly worried about the subject's reputation as a pursuit of amateur scholars or a loose and idiosyncratic body of gazetteer knowledge associated with adventurous discoveries of the past. Ironically, many geographers ascribed this "common" reputation of geography to the longstanding success of Jedediah Morse's *American Universal Geography*, popular throughout the early republic and antebellum periods. Through texts such as these, some believed, geography gained a reputation as "wondrous" and "encyclopedic" by the mid-nineteenth century. Though the subject's weakness cannot be attributed to a few authors, these characterizations of geography—as a record of exploration, a folkloric compendium of exotic, distant lands, and a school subject that stressed rote memorization—did cause some to worry about, and insistently defend, its legitimacy as a university discipline.[2]

Rather than just a tool of exploration and mapping or a collection of curiosities, geography was—in the eyes of its practitioners—a structured discipline with a delineated scope and method. Geography's new challenge was to use, rather than to collect, the mass of information gained in the nineteenth century. This institutional need to legitimate academic geography directly shaped the discipline. More fundamentally, the discipline's intellectual malleability opened it up to influences beyond the university, a modern example of David N. Livingstone's argument that "the very stuff of geography, its language, it methods, its theories, and its practice, were constituted in the midst of the messy contingencies of history."[3]

The Institutional Origins of Academic Geography

Like the Royal Geographical Society, the AGS began to modernize in the late nineteenth century, spurred by a declining interest in exploration and a rising concern with the development of geography as a discipline and a school subject. Outside the AGS and the USGS, however, geography had little institutional power in the years after the Civil War, when its influence was dispersed among few universities. Among the most influential and popular geographers of the era were transplanted Europeans: Louis Agassiz and Arnold Guyot. Agassiz, appointed at Harvard in 1848, was trained in the natural sciences, and is important for his contribution to theories of glaciation in producing landforms. Guyot, appointed as professor of physical geography and geology at the College of New Jersey (later Princeton University) from 1854 to 1880, was noted for introducing a new concept of geography based on the ideas of Carl Ritter. To Guyot and Ritter geography was not a descriptive catalog of the earth's elements, but an observation of the interrelationship between the land, the ocean, the atmosphere, and humans, all of which interacted harmoniously in a grand design. Though imbued with teleological motives and explanations, Guyot's work was important for introducing causal relationships between humans and their environment, which moved geography beyond description toward interpretation.

In Britain, Mary Somerville's *Physical Geography* also attempted to link human life to geographic generalization. Most of Somerville's work described the natural world, yet she was careful to make clear that humans affected their material world more than the reverse, and that race was more influential than natural environment. Her articulation of this relationship inspired the American geographer George Perkins Marsh to explore the same relationship in his own *Man and Nature*. Though Somerville and Marsh were immensely popular writers in Britain and America respectively, and early exponents of the broadened understanding of geography as the study of interactive relationships, their work did not directly influence the rise of American university geography, in part due to its theological bent.[4] As we will see, Guyot, despite his own theological approach, made a lasting mark by describing the physical environment as a complex interaction of human and physical life—into which basic framework Charles Darwin postulated an essential mechanism of change.

It was not until the late 1870s that modern geography began to appear in American universities, and only in 1898 was a separate department of geography established.[5] The late nineteenth century was a critical time when many American universities were either born or reorganized with more secular goals and more specialized fields of knowledge. Newly professionalized economists, psychologists, philosophers, geologists, sociologists, and historians worked feverishly to secure a place for themselves in this new context. In Britain, Mackinder led the professionalization of geography by touting its breadth and utility for a modern society.

> To the practical man, whether he aim at distinction in the State or at the amassing of wealth, [geography] is a store of invaluable information; to the student it is a stimulating basis from which to set out along a hundred special lines; to the teacher it would be an implement for the calling out of the powers of the intellect, unless indeed to that old-world class of schoolmaster who measure the disciplinary value of a subject by the repugnance with which it inspires the pupil. All this we say on the assumption of the unity of the subject.

Mackinder championed geography's potential to unify the study of man and nature, arguing that the subject's "inherent breadth and manysidedness should be claimed as its chief merit." But he also warned that in this new disciplinary context what geographers regarded as broad could also be characterized as idiosyncratic, rendering the subject "suspect" according to the new class of specialists.[6] Mackinder's warning reflects the odd situation of geography in both Britain and America in the late nineteenth century, where academic disciplines gained legitimacy not by all-inclusive claims of scope but by delimiting their boundaries. Yet doubts about geography's worth as a science kept it closely tied, and frequently subordinate, to geology. This relationship developed in a similar manner in the United States, and had implications for the maturation and independence of geography in the American university.

In fact, Mackinder's counterpart in the United States—the most important figure in the early professionalization of geography—was not a geographer but a geologist, a reflection of the absence of doctoral programs in geography at the time. William Morris Davis was trained in geology by Nathaniel Southgate Shaler, then appointed assistant professor of physical geography at Harvard in 1885. Davis reasoned that if geography were to mature it must first address what Mackinder had noticed as its thorny reputation as the "mother of all sciences." In a climate where specialization was the key to institutional success, this claim only hindered progress. As

Andrew Kirby has argued, while geographers considered their discipline's breadth and synthetic nature an indication of their exceptional status, other academics considered it evidence that geography lacked disciplinary coherence.[7] In fact, the skepticism of nongeographers, together with Davis's own battle to separate geography both from other university sciences and from the "common" geographical societies highlight a longstanding crisis: how does a field that is essentially synthetic explain its importance to a body of sciences that are premised on the uniqueness of their character? In other words, what would keep geography independent when it borrowed so heavily from geology, anthropology, botany, and other neighboring disciplines? Davis argued that united and firm declarations of geography's intellectual and institutional boundaries would solve this dilemma, and that assertions of breadth only encouraged incoherence.

Shaler and Davis created a formal course of training in physical geography—the study of the surface features of the earth—and mentored the first generation of trained geographers in the United States. Though Harvard was not the only center for graduate training in the late nineteenth century, Davis's presence made it the center for physical geography, which at that time was the most important component of the young discipline.[8] As a geologist, Davis was oriented toward the physical rather than the human elements of the landscape, and during the 1880s and 1890s he advanced an idea that applied Darwinian evolution to the study of the earth. The resulting science of geomorphology would interpret rather than simply describe the physical landscape and frame the world as dynamic rather than static. Though his ideas were not revolutionary—he freely admitted the debt he owed to others—Davis hypothesized how different elements of the environment interact to produce change over time through concepts such as the cycle of erosion. As chapter 5 suggests, the curriculum reforms of the 1890s incorporated this idea of geography as an evolutionary science, and Davis's contributions led to the decision to make physical geography a college entrance requirement in 1893. Davis's concept of geomorphology helped legitimate geography in the university and gave geographers a tremendous boost in the process.[9] But if geomorphology gained for geography some measure of clarity, it also reinforced geography's identity as a subfield of geology.

Davis tried to emancipate geography institutionally as well. While other sciences and social sciences gained legitimacy as narrow, rigorous intellectual endeavors with established professional societies, geography was dogged by its reputation as a "genteel social activity." Many since Davis have argued that the reasons for geography's slow academic growth were its reputation as overly accessible and overrun with amateurs in organizations such as the AGS, the NGS, the Geographical Society of Philadelphia, and the Appalachian Mountain Club.[10] These groups were endlessly irritating to Davis and others because they reinforced the sense both outside and inside the academic community that geography was either a utilitarian tool for surveyors or a curious leisure pastime of travelers. In fact, the belief that armchair geographers damaged the professionals motivated much of Davis's behavior. Not only did he dissociate himself and his cadre from these organizations; he also attempted a coup at the National Geographic Society in 1900 in hopes of returning the *National Geographic* to its original scholarly form. To Davis, the Society represented the troubling persistence of an amateur tradition in the midst of a struggle for professional recognition. As he speculated, "it can hardly be doubted that the acceptance of low standards for membership in our geographic societies has had much to do with the prevailing indifference regarding the development of a high standard for the qualification of geographical experts." Davis was not alone in his concerns about the organizational state of American geography. The National Geographic Society, confident after its experience during the Spanish-American War, effectively ignored the professional geographers by inviting the International Geographic Congress to hold its 1904 meeting in America. When the offer was accepted, many academic geographers shuddered at the prospect of the comparatively amateur National Geographic Society hosting a conference of the most esteemed European geographers. This, they feared, would only confirm assumptions among scientists that American geography was stunted both intellectually and professionally. In the pages of *Science*, academic scientists such as Israel Russell warned that "we are not prepared for such an invitation," and hoped that a "real" society could be in place well before the IGC in 1904.[11] Davis lamented the sorry state of geographical societies around the country, and encouraged a new organization that would be closed "to the mere traveller, the lover of outdoor nature," and the dilettante. Like other newly professionalized scientists before them, geographers wanted an organization to distinguish them from their nineteenth-century counterparts — amateurs and government scientists.

With this goal in mind Davis helped found the Association of American Geographers in 1904, a new entry in the growing list of associations formed around academic disciplines. The association would maintain high standards for admission and give geographers a forum in which geographic scholarship could be considered seriously and the discipline's future freely contemplated.[12] Among the organization's basic tenets was exclusivity, in both membership and mission, for as Davis argued, "geography will never gain the disciplinary quality that is so profitable in other subjects until it is as jealously guarded from the intrusion of irrelevant items as is physics or geometry or Latin."[13] Geologists were initially welcomed into the society in order to strengthen it numerically, but by the 1910s these prospective members were generally deferred to preserve the disciplinary purity that had become so important to geographers. As the century opened, then, geography's place in the university looked relatively bright. With a professional association closed to rival scientists and armchair explorers, and with Davis's intellectual leadership, geographers quietly joined with the NGS to cohost the Eighth International Geographic Congress in Washington, D.C. A conference that had previously been held in Europe's most cosmopolitan cities, the event symbolically announced the debut of American academic geography in 1904.

A TURN TOWARD THE HUMAN

By the turn of the century Davis had trained many young geographers in physiography, but the pressure to develop apart from geology encouraged both him and his students to broaden their scope in order to study what geology did not — the organic elements of the landscape. Initially Davis and Shaler imagined this would include all aspects of life, but soon they began to encourage their students to concentrate on the *human* dimension of organic life. Davis, Shaler, and their students further refined this subfield as the human *response* to the physical landscape. This approach, which Davis called ontography, appeared to offer geography a reliable path to intellectual and institutional success; it was theoretically grounded yet — unlike physiography — also gave geographers an identity apart from geologists. By carving out a scientific relationship of cause and effect ontography became the "flagstone of geography's claim to a niche in the evolving division of scientific labor."[14] Physiography and ontography therefore shared the spotlight at the turn of the century: while the first causally linked elements of the environment with each other, the second linked the environment with its human inhabitants.

Though the center of these intellectual developments was unquestionably Harvard, by the turn of the century graduate work in geography could be found at the University of Chicago, the University of Pennsylvania, and Yale University. And at each institution geography was taking on a decidedly human cast, though some students continued to concentrate on the physiographic and geologic elements. Harvard's earliest rival in geography in the early twentieth century was the University of Chicago. Opened in 1892, the University had from the outset a dynamic geography teacher in Rollin Salisbury, who, like Davis, had been trained in geology. In 1903 he convinced Chicago to establish the first separate department of geography in the United States to offer graduate degrees. His popularity—and that of the department—grew immediately, even moreso after he began to train a generation that went on to influence geography in colleges and universities around the country. In fact, from 1903 to 1923 Chicago was the only university offering a doctorate in a discrete department of geography.[15] At the University of Pennsylvania, geography originated as a byproduct of the government's efforts to build an isthmian canal in Latin America. Two economists, asked to predict the canal's implications for trade, realized that such questions demanded geographic research. As a result, Emory Johnson and his student, J. Russell Smith, founded the Department of Geography within the University's Wharton School of Finance. Thereafter, Pennsylvania's geographically inclined economists encouraged the human aspects of the study, training influential textbook writers and teachers throughout the early twentieth century.[16]

Though the genesis of geography at Harvard was very different from its incarnation at Pennsylvania, both were premised on a causal relationship between human behavior and geographic circumstance. In fact, most early geographers conceived of their discipline as having exceptional power to bridge the natural and human sciences. From the mid-1890s to the First World War the prospect of uniting nature and culture through geography seemed both feasible and imminent at Pennsylvania, Chicago, Yale, and Harvard. As a result, theories of influence and causation that united these two spheres held special appeal for geographers.

Jurgen Herbst has argued that Darwinian concepts of evolution helped geography mature from its encyclopedic, fact-based form into a rational and dynamic study of life. Geographers used natural selection to explain the interaction between physical environment and life as one of inorganic control and organic response, and in the process they found their scientific principle. When geographers widened this interactive framework to include human behavior in the category of organic response, the range of ideas known loosely as Social Darwinism became pivotal. Geography—defined as nature's influence upon human behavior—became a study of why some groups struggled and survived in their environments while others languished.

Herbst explained that the pressures on geographers to prove their worth in the university only intensified their reliance upon causal, potentially deterministic frameworks: without a theory to unite man and land, the studies under geography arguably belonged to other disciplines.[17] More recently, David Stoddart has argued that Darwinian evolution clearly influenced the discipline of geography, although it was misinterpreted by American geographers. Specifically, geographers neglected the idea of random variation and replaced it with a continuous process of change. Furthermore, the Darwinian element of struggle was exaggerated and accelerated to incorporate humans into the ecological world: what Darwin understood as a long-term process became an observable phenomenon for American geographers, and what would otherwise have been considered discrete and unrelated data became unified under theories of environmental influence.[18]

Other scholars have convincingly argued that the intellectual roots of American geography owe less to Darwin than to his predecessor, Jean Baptiste de Lamarck. Lamarck understood evolution in more flexible and less deterministic ways than Darwin by suggesting that characteristics acquired through life could be passed biologically to offspring. While Darwin's ideas were adapted from the natural to the human world, Lamarck explicitly designed his theories of heritability to explain human development. As George Stocking has observed, Lamarckian thought and modern behavioral science were well suited to one another in the late nineteenth century. Just as "the social sciences explain the behavior of man as the product of his interaction with his social environment[,] Lamarck made the organism's behavioral responses to environmental changes the mechanism of physical evolution."[19] In other words, Lamarck brought the study of nature and culture together by linking environment, biology, and social progress. Pioneers in American natural and social science such as Auguste Comte, Lewis Henry Morgan, Herbert Spencer, John Wesley Powell, Lester Frank Ward, and G. Stanley Hall all accepted the heritability of acquired characteristics. As Herbert Spencer wrote:

> it needs only to contrast national characters to see that mental peculiarities caused by habit become hereditary. We know that there are warlike, peaceful, nomadic, maritime, hunting, commercial, races — races that are independent or slavish, active or slothful; we know that many of these, if not all, have a common origin; and hence it is inferable that these varieties of disposition, which have evident relations to modes of life, have been gradually produced in the course of generations.[20]

The peak of Lamarckian influence over the American behavioral sciences came in the 1890s; thereafter the rediscovery of Mendel's laws in 1900 dealt a punishing blow to its credibility. In the case of geography, neo-Lamarckian explanation peaked slightly later and survived much longer. Livingstone suggests that Lamarckism offered geography what it offered the behavioral sciences generally: a way to join studies of humanity with the natural environment. Institutionally, geographers were drawn to

Lamarckian constructions because they helped define for geography a scope apart from the human sciences by incorporating the influence of the physical world. At the same time, and somewhat paradoxically, Lamarckian explanations defined geography apart from geology, thereby strengthening the former's independence in the modern university.

Lamarck also answered many of geography's intellectual dilemmas, and Livingstone has outlined the sources of this debt.[21] Lamarck's basic premises—that acquired characteristics are then inherited, and that the conscious use of an organ increases its capacity and leads to evolutionary progress while disuse results in degeneration—emphasized the drive to adapt to one's environment. This in turn placed psychological and behavioral processes at the center of observation. The focus on the conscious adaptation to one's physical environment removed the random chance of Darwinian evolution and allowed the strength of an individual, a culture, a race, or a nation to be passed on. Fate and natural selection were replaced with responsibility and intent. While Darwin granted nature control over human development, Lamarck gave humans some power over their destiny. Consciousness and will were needed to survive and prosper: good character was acquired by conquering one's physical environment, while bad character resulted from the latter's victory over human intent and ability. Furthermore, the pace of evolution, which spanned thousands of years in Darwin's scheme, could be observed over the course of a few generations under Lamarck. This was especially appealing to geographers since it made the environment's influence on human behavior an observable phenomenon. Put another way, Lamarck created a process for geographers to investigate.

Lamarckian theories of the human-environment relationship also added another highly attractive dimension to geography. The premise of Lamarckism—the inheritance of acquired characteristics—accelerated the pace of change. By placing some control in the hands of its subjects, Lamarckism made for an open-ended theory of human behavior that allowed geographers to grant their subjects as much autonomy as necessary. The nature of this biological mechanism, however, was vague, which made for diverse—even chaotic—methodologies among geographers. Indeed, the terms "race" and "nation" became conflated under Social Lamarckism, denoting both cultural identity and biological constitution and blurring the distinction between evolution and history. Such vague differentiation meant that "geographers with Lamarckian inclinations" moved easily between biological and social explanations.[22] For geographers, Lamarckism was more often than not a theory of human *response* to the environment. Implicit in this formulation was the assumption that humans were by-products of the earth, but also in some measure responsible for their interaction with the environment. By the early twentieth century, the reaction to the physical landscape—more than the landscape itself—was what preoccupied geographers. And this theme of response to the environment contained a powerful moral dimension of responsibility that nicely suited the didactic purposes of school geography.[23]

This is not to say that geographers recognized the influence of Lamarck over their study, as many were unacquainted with his work. But throughout the first half of the twentieth century geographers did write about the interaction of humans and their environment with firmly Lamarckian presumptions. These assumptions were not always conceived in deterministic ways. While some geographers invoked them as evidence of an intellectual and social hierarchy in order to justify American expansionism or European imperialism, others used them to open up possibilities for social change.[24] It was just this flexibility—or indeterminacy—implicit in Lamarckism that allowed it to shape geography long after it had been discredited in other behavioral sciences. In fact the range of interpretations possible in Lamarckian expositions is precisely what made it attractive to geographers. This flexibility also encouraged geographers to design explanations of human activity on an epic scale, though such explanations became increasingly rare as the age of specialization wore on.

By briefly profiling a few of the first professional geographers we can see how school and popular geography developed out of this intellectual tradition. In the first generation these men included William Morris Davis, Halford Mackinder, Friedrich Ratzel, and, to a lesser extent, Rollin Salisbury. In the second—the first to be trained as geographers in the new universities—the most prominent were Ellen Semple, Ellsworth Huntington, Albert Perry Brigham, Richard Elwood Dodge, and Isaiah Bowman. Two others, Emory Johnson and J. Russell Smith, also fit this intellectual profile, and were among the most active practicing geographers despite their formal training as economists. Of the two, only Smith will be considered here and in chapter 5 for his legacy as an author and a mentor.

What we find is that these geographers assimilated the goals of American expansionism into the very foundation of their discipline. That is, Davis and his students accepted a Lamarckian form of environmentalism as the mechanism for human geography. The influence of Mackinder, as well as that of Germany's leading geographer, Friedrich Ratzel, is instructive here. Trained primarily in zoology, Ratzel posited a relationship between human history and physical geography much like Davis's relationship between inorganic development and organic response, calling it anthropogeography. But where Davis was tentative, Ratzel was bold. Ratzel applied concepts of Darwinian struggle to human society in order to frame the state—not the individual—as an organism that was forced to expand in order to survive. Like those of Turner, Davis, and Mackinder, Ratzel's ideas allowed geographers to link nature and culture. And, like Mackinder and Turner, Ratzel developed an approach that fit well with the contemporary expansionist rhetoric of Josiah Strong, Alfred Thayer Mahan, and Theodore Roosevelt, each of whom regarded the nation's future as thoroughly international. In 1900 Ratzel published *The Sea as a Source of the Greatness of a People*, a treatise that owed more than a little to Mahan's *Influence of Sea Power upon History*. Both emphasized sea power as central to national survival in the twentieth century.[25]

Mackinder also held strong views about environmental influence that he hoped might allow geographers to forge a new identity. As argued above, he saw the descriptive character of geographical knowledge giving way to one premised on the interaction between man and the land, which naturally led him toward theories of environmental influence. In the process, Mackinder hoped, geography would gain theoretical, practical, and intellectual grounding, making it useful to the historian and the scientist, the merchant, the statesman, the teacher, and the student. In 1904 Mackinder published "The Geographic Pivot of History," which gave him a sterling reputation—even more in the United States than in his native Britain—as a pioneer in both geographical thought and its application to national policy. In this article Mackinder articulated the geopolitical dimensions of international relations, premised on the assertion that the end of the century marked the end of an era dominated by exploration. In the new age, the true test would be manipulating the knowledge made available by exploration in order to develop territorial strategies. Mackinder assumed that the crucial moment in historical change was the human response to the environment—in other words, how individuals and societies chose to apply knowledge to the conditions before them. Through this dynamic, the historical became intertwined with the geographical, transforming political geography from a recitation of boundary limits and capital cities into an interpretive survey of modern nation-states based on their position, resources, and diplomatic relations.[26] This scope made environmental influence central to the development of twentieth-century American geography. Mackinder's theory of history, like that of Turner, was based on the idea that the human experience of geographical space had changed in fundamental ways in the late nineteenth century. As Livingstone has noticed, Lamarckian influence undergirded the geographical ideas of both Ratzel and Mackinder, because both assumed that the impetus for change lay primarily in external environmental circumstances rather than in genetic constitution.

This marriage of geographical influence and human response directly fed the budding discourses of geopolitics at the turn of the century. As Stephen Kern has argued, the rise of geopolitics also owed much to the cultural and technological changes that took place from 1880 to 1914. Standardized time, the advent of flight, the expansion of railroads, and advances in radio and communication all revolutionized the common experience of space and time. Kern argues that Ratzel and Mackinder used geopolitics to come to terms with the changed sense of distance that resulted from these innovations. Thus it should be no surprise that in the 1940s, when another reorientation of space was brought on by the air age and a global war, Mackinder's thought was resuscitated and used as the basis for modern geopolitics (which will be discussed further in chapter 6).[27] Mackinder's "Geographical Pivot" was among the most controversial of these new geopolitical ideas, with its contention that the sea age was coming to a close, being replaced by an age of land power that made railways crucial in the development of trade. With the ideas of Mackinder premised on a sort of geographic causation, and with Turner's ideas of the frontier close at hand as well, geopolitics gained legitimacy as a shrewd interpretation of the relationship between the size and location of a state on the one hand, and its politics and history on the other. Friedrich Ratzel's own geopolitical project was broader; he intended to study the relationship between the earth and human history. But as Kern notes, Turner, Ratzel, and Mackinder all worked to connect human history and physical geography, symbolizing the larger mission of academic geography to unite man and land. Davis was beginning to move toward this ontographic theory of geography by the end of the century, though he avoided its geopolitical interpretation. Though he had not completely worked out its implications, he used it to influence many of his students at Harvard, who would go on to shape the future of American geography.

THE SUCCESSORS

It was a dramatic and complicated shift from the gazetteer form of geography in the nineteenth century to the dynamic, interpretive, and explanatory strains that came to prominence in the twentieth. Much of the impetus for this shift came from the institutional pressures within the university, but the actual transition developed in the context of expansionism. Almost all of the best known students of Davis and Ratzel at some time articulated ideas of environmental influence in the hopes of uniting human history and physiography, and all confronted at some point the premise of Lamarckian or Darwinian thought. They did so as much as anything because of institutional imperatives in their field. The need to claim a niche in the ever more specialized university prompted them to seek intellectual approaches that gave their discipline a claim to uniqueness among neighboring natural and social sciences.

The geopolitical ideas of Frederich Ratzel crossed the Atlantic with the first American woman to work toward an advanced degree in geography and the only female charter member of the AAG. Ellen Churchill Semple introduced anthropogeography where physical geography had dominated, shifting American geography from a natural toward a human science. Semple was not the first to frame geography as a human study, but it was she who articulated, in unambiguous language, the relationship between human behavior and environmental conditions.[28] Semple married Ratzel's methods to Turner's concerns by using environmental determinism to explain American history.[29] Ten years after Turner set forth his theory of the American frontier, Semple wrote *American History and Its Geographic Conditions*, delineating the influence of the latter on the former. In this and her next work, *Influences of Geographic Environment*, Semple took her cues from Ratzel, arguing that living organisms evolved from simple to more complex forms through adaptation to the physical environment. The larger a state, a race, or a people, the better their chance of survival relative to others competing for the same resources.

> The barrier of the Atlantic was a basis of natural selection among the early colonists, so that in general only the fittest, the robust and enterprising, reached American shores. The abundance of opportunity in this virgin land constantly attracted foreign immigration to reinforce the American population. Easy conditions of earning a livelihood induced early marriages and the consequent large families which became a factor in the expansion of the nation. When with growth of population competition grew stronger and land scarcer, with a fine impatience of these altered economic conditions the dweller of the frontier moved westward to take up unexhausted lands, an unused "range," and also to seek that unconstrained life of the backwoods which has a powerful charm to the natural man. Thus fullness of opportunity bred the migratory instinct, and geographical conditions favored it.[30]

Semple's premise was implicitly Lamarckian. Though she frequently credited the nation's success to the superior pool of immigrants, throughout the narrative she imbued the environment with power to make history and remake individual physical, intellectual, and moral constitutions. In this regard Semple was perhaps the most deterministic of geographers, but her work won her the praise of more possibilistic colleagues, even Turner himself.[31]

The writings and teachings of Friedrich Ratzel also influenced the work of J. Russell Smith, who studied with Ratzel in 1901 and received his doctorate in economic geography from the University of Pennsylvania's Wharton School of Finance. Smith had come to geography from economics, convinced after his work with the Isthmian Canal Commission in 1900 that a grasp of geography was necessary for the future growth of American trade.[32] Smith's deterministic focus on the influence of climate over human economic behavior indelibly shaped primary and secondary school geographies in the early twentieth century, as argued in chapter 5. His lifelong interest in economic geography also propelled him to create a geographical subfield in the Wharton School in 1906, and to help found Columbia University's geography program after World War I, also within the School of Business. Smith was notable as one of many geographers who eventually repudiated environmental determinism for more dynamic approaches.

Aside from Semple and Smith, most other influential geographers of this period came under Davis's training at some point in the 1890s or early 1900s.[33] Richard Elwood Dodge studied under both Davis and Shaler at Harvard, and, along with Semple and Albert Perry Brigham, presided over the conceptual shift from a physiographic to a human-centered geography. Dodge also founded the *Journal of School Geography*—later renamed *The Journal of Geography*—and edited it from 1897 to 1910, solidifying his place as a leader in teacher training and school geography. After leaving Harvard, Dodge was one of the first to teach the "new" geography of human-environmental relations at Columbia Teacher's College; he taught there from 1897 to 1916. Under Dodge's guidance at Columbia, courses in commercial and industrial geography began to appear in 1905 and 1911 respectively, moving physical geography to a secondary role in the curriculum.[34] As we will see in chapter 5, Dodge's schoolbooks used the causal relationship to bring humans and their environment together so that geography might link the social and scientific subjects. Though initially taken in by the elegance of determinism, he was able to avoid its doctrinaire implications and reach a position that allowed for ambiguity where Semple had been fairly rigid. To Dodge, the physical environment was conceived as offering opportunities as well as setting limits on individual choices. In avoiding determinism and insisting on a more human and less physical interpretation of geography, Dodge was one of the earliest exponents of geography as the ecology of human society.

Albert Perry Brigham, Dodge's fellow pupil, was a former minister who also became a disciple of Davis before teaching at Colgate University from 1892 to 1925. Like Semple, Brigham was drawn to the influence of environment upon history, crystallizing his own theory of this relationship in his *Geographic Influences in American History* (1903). Brigham was less dogmatic about the influence of the natural environment, giving decisive weight instead to the determinism of race. Racial tendencies, he wrote, "whatever their source, are always obscuring or swerving the lines drawn by nature." Nevertheless, the sheer abundance and topographical expanse of the New World made it hospitable to growth and development, factors that were fully utilized by the superior breeds who managed to settle the land. On the Englishman, Brigham argued: "Not because of geographic opportunity, but by virtue of qualities that inhered in his race, he reached northward and southward, dispossessing his Latin neighbors, and poured at length across the barrier, and swept the valley of the Mississippi."[35] While celebrating the increasing diversity of the American population through the nineteenth century, Brigham cautioned against the influx of more recent groups from south and southeastern Europe. The geographic abundance of North America could not alone ensure its continued prosperity—care had to be taken in developing its national stock as well.

Brigham's ideas about environmental and racial influence over human behavior were put to work in defense of American expansionism at the turn of the century. In Brigham's geography, the Ratzellian imperative for the state to expand merged beautifully with the peculiar suitability of the Anglo-Saxon and Teutonic races to lead. Brigham's own ideas of American exceptionalism functioned in this mix to blur the line between history and destiny. Recent gains from the Spanish-American War, together with the newly discovered riches of Alaska and the commercial advantages anticipated with the opening of the Panama Canal, would guarantee American supremacy abroad. While racial superiority and manifest destiny helped explain America's rise to power, the abundance of the environment would ensure its continuing dominance.

> The geographic conditions for American growth seem to have been perfect. At a critical time in the history of European thought and life, a sturdy people needed a new field. That field was opened to them by the voyagers of the fifteenth and sixteenth centuries. . . . The very largeness of the American problems has helped to make a people able to solve them, and that people now finds itself fronting the two great oceans, where, more easily than any other nation, it can reach out and touch every part of the world. These conditions, in their entirety, are unique in history. They are largely

geographic in their character, and they only need the perennial support of the basal moral qualities to insure to our country un-failing leadership among the nations.[36]

These excerpts illustrate Brigham's conditional use of determinism. For him it was a question of weighing different influences rather than assigning total control to the physical environment. Later, however, he consciously avoided causally connecting human behavior to the environment, and became one of the more vocal critics of geographic determinism. Though frequently criticized as being unscientific and having disastrous consequences for geography's reputation among the social sciences, the link between environment—especially climate—and behavior became one of the most enduring influences on twentieth-century school geography. As argued in the next chapter, Brigham was one of the primary conduits through which environmentalist and racialist strains were brought into American geography schoolbooks.[37]

One of the most popular of Davis's students pursuing the influence of climate over human behavior was Ellsworth Huntington, who brought this human dimension of geography to his tenure at Yale, first as a teacher from 1907 to 1915 and then as a research associate from 1919 to 1945. For Huntington, geography's initial focus on the physical environment provided a natural grounding for its explanation of social relations. Huntington's fieldwork in Asia during 1905 and 1906 led him to posit a relationship between climate and history. First and foremost was the premise that climate fluctuated over the long term. That certain eras had been wetter, drier, cooler, or warmer than others had already been established; what Huntington asked was how these climatic trends shaped the habits and the character of human societies. In this framework, climate became the stuff of history.[38] Huntington's *The Pulse of Asia* was widely hailed in the early twentieth century, rivaling in popularity the works of both Semple and Brigham on American history and geography. The critical reception of Huntington's work was quite positive, and few geographers were troubled by its deterministic overtones at the time of its publication.[39] During the 1900s and 1910s, Huntington was by far the most vocal exponent of this causal relationship, pointing to climate as the strongest influence over human behavior. In subsequent writings, Huntington showed no hint of modifying this stance. The preface to *Civilization and Climate* argued that one of geography's primary functions was to ascertain the dependence of human character on geographic environment. As Huntington wrote, we acknowledge our "faith in climate" constantly, if implicitly, by talking about it, changing our location relative to the seasons, and responding emotionally to given trends in the weather. "Yet, in spite of this universal recognition of the importance of climate, we rarely assign to it a foremost place as a condition of civilization."[40] Huntington devoted his life's work — through hundreds of articles and a slew of monographs — to closing this gap between common sense and scientific inquiry.

Huntington argued that civilization would flourish only in areas of "stimulating" climate, and falter in areas of extreme heat — particularly the tropics. It was, he contended, a complex mix of racial aptitude and environment that ensured survival and prosperity, but even racial constitution had been made by environment, since "many of the great nations of antiquity appear to have risen or fallen in harmony with favorable or unfavorable conditions of climate." Here again notice the conflation of race and nation, a common theme in many areas of geographical knowledge (figs. 4.1 and 4.2).[41] Like many of his contemporaries, Huntington's claims about the influence of climate on human behavior rested on Lamarckian assumptions. In the introduction to *World-Power and Evolution*, published in 1919, Huntington emphasized the role of Darwin in his work, but included an important — Lamarckian — condition:

> The sum and substance of biology is evolution, the Darwinian idea that no type of living creature is permanent. . . . Variations occur, and natural selection by means of the environment ruthlessly exterminates some of them and preserves others to form new species. The variations are possibly sudden and marked rather than gradual and slight as Darwin supposed, but that does not alter the main idea.

Later he referred to the biological heritability of acquired characteristics.

> No one will question that our efforts to train the next generation in the right way must be redoubled. . . . we must give tenfold or a hundred-fold greater weight to the great problem of eugenics. Our country's children must have a good *inheritance*. The best inheritance and the finest training . . . [require] *health*. How many human ills arise because well-trained people with a good inheritance fail to do their part through ill health or nervousness?[42]

Huntington believed that external circumstances — such as weather — could not only modify behavior but also cause mutations that would then be passed on. This is not to say that Huntington did not appeal to race to explain human diversity — the opening chapter of *Civilization and Climate* makes quite clear his belief in a hierarchy of racial capacity. But he was also

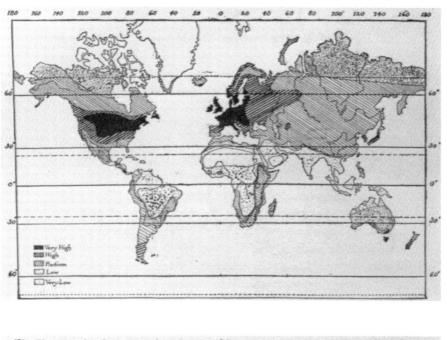

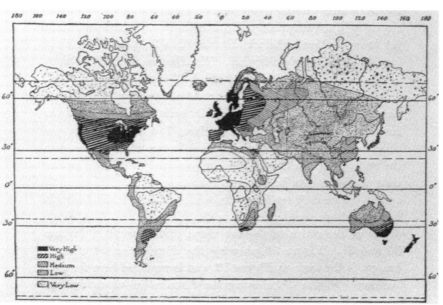

Figures 4.1 and 4.2 Two of the most widely reproduced of Huntington's maps, depicting the distribution of human energy on the basis of climate (upper) and the distribution of civilization (lower).

optimistic about the power of environment to reformulate our mental and physical strength.[43]

Huntington's controversial studies enjoyed widespread recognition, or at least notoriety, among geographers and social scientists. *Civilization and Climate* was revised and reprinted four times within ten years. The climatological essays he published in *Harper's Magazine*—"Work and Weather," "Climate and Civilization," and "Is Civilization Determined by Climate?" —inspired anger, curiosity, and enthusiastic support from readers.[44] But by the early 1910s, historians and geographers increasingly criticized Huntington's use of scattered physiographical data, his highly subjective definitions of "civilization," and his deterministic language. The extent of Huntington's determinism even made him doubt the legitimacy of other social sciences. In declining an invitation to appear on a panel with Semple at the 1916 meeting of the American Historical Association, he wrote, "I suppose that I may be over-sanguine, but I am persuaded that within the next generation the philosophy of history will suffer modification because of a new understanding of the effect of physical environment, especially climate, upon human energy."[45] Huntington continued to search for environmental influences over human development until his death in 1945, and this preoccupation left him isolated from a profession increasingly wary of environmentalism in any form. In the interwar era he was among the only leading American geographers to claim environmental influence as the key to a unified conception of geography. His popularity outside the university, however, remained high throughout this period.

The best known of Davis's students—and among the most influential geographers of the twentieth century—was Isaiah Bowman. A student of Davis in the early 1900s, Bowman incorporated a strong sense of physical geography into his own version of the discipline, and until his death remained suspicious of human geographers who slighted the importance of basic physiographical principles. Yet his own mechanism of the relationship between human and physical geography was far from that of Huntington. After studying under Davis, Bowman taught at Yale until 1915. During his tenure there, which coincided with that of Huntington and Hiram Bingham, Yale was considered to have one of the best departments in human geography. The department deteriorated after Bowman and Huntington left and was eliminated during World War I.[46] In 1915 Bowman took over the directorship of the highly prestigious American Geographical Society, where he remained until 1935. It was in this position that Bowman became perhaps one of the most influential geographers of the early twentieth century. One of his goals had been to make the Society more relevant to social and political problems, and by placing its resources at the disposal of the federal government he ensured that the Society's vast reserve of maps would be pivotal in the postwar reconstruction of Europe. The entire experience put the AGS in higher standing and closer relations with the government, enabling the former to take on increasingly ambitious tasks after the war.[47]

Furthermore, as Director of the AGS Bowman assisted the government in its preparations for the Paris Peace Conference, primarily through work in the Inquiry Committee. Woodrow Wilson established the Inquiry Committee five months after the United States entered the war, and designed it as an information-gathering body that would eventually aid the postwar peace in Europe. Organized under Colonel House, the Committee was composed of historians, geographers, journalists, economists, and regional specialists, most of whom were drawn from universities. At different times, it was led by Sidney Mezes (president of City College of New York), Walter Lippmann, and Bowman.[48] Altogether, the 150 members of the Committee produced 2,000 reports and documents, and 1,200 maps, examining nearly every area of the globe but with particular attention to the political boundaries in Europe. Bowman led this work for a year before the armistice, and was instrumental in Paris as the liaison between the cartographic experts and their superiors. The Committee's cartographic work was essential to its mission, and thus the group produced every conceivable kind of map—political, ethnic, historical, and economic.[49]

It was in Paris after the war that Bowman met some of France's leading geographers, notable for their ability to construct alternatives to geographical environmentalism that emphasized human will and reversed the causal relationship. These encounters, plus his firsthand experience with the destruction wrought by war, underscored to Bowman the importance of economic and political explanations of human behavior, and conversely illustrated the limitations of a geography based on environmentalist frameworks.

After the war Bowman shifted his attention toward this new interest in political and economic geography, and by extension toward more fluid and open-ended interpretations of the human-environment relation. This new interest was first articulated in *The New World*, his survey of international relations. In this study Bowman described a world transformed by war, one where the United States would play a central role in the development and management of its own and other economies. Like the early writings of Walter Lippmann, Bowman's *New World* was optimistic about the capacity of the public to make informed choices about the direction of international relations. As some have recently argued, this treatise was an early statement of liberal internationalism, and because it came well before its time, is a uniquely candid articulation of American interests in the world. Coming on the heels of his work in Wilson's administration, *The New World* solidified Bowman's reputation outside the profession as a leader in the geographic dimensions of international relations.

Bowman's movement away from environmental determinism signaled a shift in the postwar development of American academic geography. All the figures discussed in this chapter had at one point framed geography as a subject that causally united human behavior and the physical environment. It was in large part the institutional position of the discipline that encouraged these intellectual developments, properly seen as attempts to link two disparate sets of geographical data under a single interpretive framework. What Livingstone has termed "the geographical experiment" peaked in the early twentieth century and fueled a range of approaches linking humans to their natural environment, such as climatology, commercial geography, and geopolitics. When geographers were asked in 1914 what the primary responsibility of the discipline was, twenty-two of the twenty-nine respondents listed "the exact determination of the influence of geographic environment." Next in popularity was the definition of geography as regional differentiation, which would grow after problems of environmental influence were discredited in the 1920s and 1930s. The two least cited tasks were "exploration of unknown or little-known places" and the study of physical geography, which indicates the degree to which geography had shed its nineteenth-century identity.

Yet if there was relatively broad interest in the problem of "geographic influence," this still masked a disturbing level of disorganization in the discipline that would come back to haunt it. In fact, the aforementioned poll of geographers in 1914 opened with the statement that "there prevails a general impression, even among geographers themselves, that there is little or no agreement as to what geography is or what its purposes and problems are." Eight geographers admitted that the discipline's primary problem was the lack of a coherent scientific relationship, which suggests that such a young field could not easily divide its labors between the physical landscape and the human response to this landscape.[50]

The state of geography in 1914 might be likened to a calm before the storm. Geographers were comfortable explaining human behavior as a function of physical environment, but after World War I assumptions about the hierarchy of progress and the supremacy of European civilization seemed dubious. As social scientists had argued throughout the 1910s, it was not nature but culture that would be the new focus of social science, evidenced by the rapid growth of sociology and the decreasing power of Social Darwinism as an explanation for human behavior. These developments forced geographers to search for a new conceptual framework for their discipline, one they hoped would continue to unite the disparate elements of their field. Geographers came to diverse conclusions about the future of their discipline, but many agreed that it owed as much, if not more, to the influence of humans as to the study of landscape. Concurrently, geographers began to feel the institutional pressure to specialize, and to throw off the expansive breadth that had been their source of pride in the late nineteenth century. Their response to the decline of environmentalism was to shrink their vision of geography, narrowing their study in both area and topic.

Yet in the process interwar geographers distorted their intellectual past. As argued earlier in this chapter, environmental determinism was never unconditionally asserted by the leaders in the field. Instead geographic influence was often weighted as the most important of many factors, including cultural traditions. After the war, geographers transformed these earlier geographers, such as Semple, Huntington, and Brigham, into apostles for outright determinism. As much as anything they did this to demonstrate their own credibility as avowedly antideterminist. These repudiations of geography's "deterministic" past created a substantial void in the discipline, and the interwar period saw many attempts to fill this vacuum. For all their damaging implications, causal constructions—however unsystematic—had been unparalleled in their ability to integrate physical science and human behavior. Discrediting these models created a struggle for disciplinary authority, accompanied by the rise of regional rivalries among university departments and "schools" of geographic thought. One of the century's most influential geographers, Carl Sauer, characterized the interwar period as "the great retreat." As we will see in chapters 5 and 6, this "retreat" had very real consequences for the development of geography in American schools. This enduring conceptual challenge of linking nature and culture—man and land—was central to the intellectual development of school geography throughout the late nineteenth and early twentieth centuries. And just as the turn of the century witnessed the attempt by university geographers to modernize their work by transforming it from a technical subject into a dynamic science, so too was it a period of modernization for school geography.

5

School Geography, the "Mother of All Sciences"
1880–1914

Geography was well suited to nineteenth-century schools, a uniquely broad subject—adventurous yet utilitarian, synthetic yet scientific, interesting yet rigorous—that held out the promise of a multifaceted, integrated study. In fact, one of the most widely used geography textbooks of the late nineteenth century opened with the following proof:

> That which teaches us the relations between the earth and man
>   must be the most useful of studies.
> Geography teaches us the relations between the earth and man.
> Therefore, Geography must be the most useful of studies.[1]

This ideal of geography's breadth would not survive into the twentieth century.

This chapter traces the modernization of geography in the American schools, just as prior chapters surveyed this change in other areas of public life: mass-market cartography, the National Geographic Society, and academic geography. The reorganization of school geography is treated last because it exhibited certain traits of—and was influenced by—each of these other traditions. Just as maps became accessible to a much wider audience in the last quarter of the nineteenth century, so too did school texts and public education. Also, like atlases and the *National Geographic*, geography as taught in the schools incorporated the nation's expansionist turn after 1898 by reconceptualizing the subject altogether. Finally, university geographers shaped school geography in two important ways before World War I. First, the enduring hope among geographers that theirs was the "mother of all sciences" shaped their approach to the school subject, especially as they assumed its centrality to the modern curriculum. Second, the tendency to frame geography as the study of human response to the physical environment became increasingly important to a subject that fell between the natural and social sciences. On the whole, school geography taught students how to understand their world according to principles of race, environment, and nationalism. After the turn of the century these explanations became explicit defenses of American commercial, territorial, and political growth.

"A Way of Traveling in Our Minds": Geography before School Reform

Though secondary school enrollments would rise sharply at the turn of the century, in the 1870s and 1880s attendance was limited. Lawrence Cremin has argued that the goals of education in this period were "order rather than freedom, work rather than play, effort rather than interest."[2] The focus on strengthening mental capacity was built on a system of memorization and recitation, which later reformers vilified as the most damaging element of nineteenth-century pedagogy. Joseph Mayer Rice, a reformer who in 1892 examined the schools of thirty-six cities, was consistently disappointed with this method. In the old schools of Massachusetts and the new ones of Minnesota he found an excessively rote curriculum that encouraged poorly trained teachers to rely on textbooks and memorization drills. In Boston Rice found a particularly egregious example in the geography lesson.

> The teacher remarked: "Newfoundland is in the North Temperate Zone; the climate is cold-temperate." She then wrote the word "Newfoundland" on the board, and the words "cold-temperate climate" under it. When these words had been written, she asked one of the pupils to give the size of Newfoundland. . . . When the pupil called upon had given the size of Newfoundland, the teacher wrote upon the board, "Size — about 365 miles wide." . . . During the whole lesson no incidents were related, — nothing but facts were mentioned.[3]

Nineteenth-century geography was particularly suited to recitation through its classification schemes of climate, geology, historical events, and the "characteristics" of civilization. In this regard, geography subsumed other subjects. The earth sciences frequently fell under its broad definition as the mother of all sciences, while history complemented its factual structure. Geography was also more commonly studied than history in the nineteenth century. Only 31,000 eighth-graders in Ohio took American history courses in 1880, while 267,000 studied geography.[4] In fact, geography itself contained a good deal of history by considering racial, social, and religious subjects in an international context that historians would soon appropriate as a separate discipline. In light of this breadth, geography was central to the nineteenth-century curriculum. An entrance examination for Jersey City High School in 1885 tested students in five subjects: geography, algebra, arithmetic, history, and grammar. In the first subject students were expected to identify the earth's axis; the equator; the principal mountain ranges in Asia, Europe, and Africa; the national capitals; the states bounding New Jersey and the nations bounding Russia; the principal sources of coffee around the world; and the location of ten international cities. Pennsylvania's West Chester High School examined its graduating class of 1872 in arithmetic, geometry, philosophy, history, and geography, and in the last subject asked students "to name the highest mountains, the five largest indentations in the coast line, the seven longest rivers and the six largest lakes in North America." To the west, students reading at the third-grade level in a two-room Illinois schoolhouse were asked to describe the Bedouin and to name all the bodies of water between Cairo, Illinois, and Cairo, Egypt.[5]

Contemporary geography textbooks reinforced this definition of the subject as a litany of relatively unsystematic but easily identifiable facts. Textbooks were among the most commonly read books in the nineteenth century, and had influence beyond the schoolroom. Notoriously all-inclusive, geography textbooks considered historical, cultural, civic, and intellectual questions within a geographical discussion of boundaries, climate, resources, and "social characteristics" such as religion, government, and history of the land's inhabitants. In this regard they were directly influenced by the nineteenth-century gazetteers described in chapter 2, breaking down the human and physical landscape into discrete facts and clear categories. The mythical place of the land in American culture was also mirrored in these textbooks. Special emphasis was placed on the size, wealth, and vacancy of the West, and recent victories—such as the Mexican War and the Oregon Settlement—were taken to be evidence of manifest destiny.[6]

One of the most popular of these textbooks was Samuel Augustus Mitchell's *System of Modern Geography*. Mitchell's text was continuously revised and reprinted—with little change—from 1839 to 1882, which itself indicates the relative consistency of nineteenth-century textbooks. Like his contemporaries, Mitchell understood geography to encompass both the natural and the human worlds. It described the physical landscape and its features, while also accounting for "the different races of men, their languages, governments, and arts, and their condition as to civilization, learning, and religion."[7] This suggests that geography possessed evaluative, if not explanatory, power. Students directly and visually confronted this approach upon opening Mitchell's text, where an elaborate frontispiece depicted the four "Stages of Society": savage, barbarous, half-civilized, and the civilized and enlightened (fig. 5.1).

Much like Cram's "Four Quarters of the Globe" described in chapter 2, Mitchell's "Five Stages of Society" divided the world according to both geographical and human categories.[8] The concept of this social, racial, and national hierarchy was influenced by Conrad Malte-Brun, a Dutch-born geographer exiled to Paris in 1800. According to Daniel Calhoun, Malte-Brun's seminal *Précis de la géographie universelle* (1810) introduced terms like "savagery," "barbarism," and "civilization" into the study of geography, loosely described as stages of cultural development people passed through. Originally Malte-Brun had used the term "half-civilized" as a synonym for "barbarian," but in the American context it became a separate category between civilization and barbarism. Calhoun also notes that William Channing Woodbridge, whose early-century geographies originated as a response to the religious and regional parochialism of Jedediah Morse's texts, added a fifth stage in 1820, "the enlightened."[9]

In America this framework quickly became a set of rigid and definitive categories by which to organize the world, and the implied distinction between stages soon became a matter of explicit order. In Mitchell's text, these five stages of society corresponded to racial and ethnic groupings, which in turn suggested the murky distinction between biology and history. Likewise, categories of race and nation were conflated much like they were in contemporary atlases and maps. These divisions of geography became a way to differentiate the world and make it complete, and to show Americans where they *fit* into a larger — often divine — order. Savagery counterbalanced civilization, and both served to illustrate the spectrum of possibility. At one corner of this schematized world students found the savage stages of "bloodthirsty, revengeful people" including the aborigines of Australia and New Guinea, and most of the Indian tribes of North and South America. Next appeared the barbarous nations of roving tribes that lived by pasturage and agriculture in Arabia, Central Africa, and Abyssinia.

Half-civilized nations — the Burmese, Siamese, Persians, Japanese, and Chinese — had a limited knowledge of agriculture and the arts, and possessed a written language, some system of laws and religion, and rudimentary trade, yet were limited by social traditions. As Mitchell wrote, "the

Figure 5.1 "Stages of Society." Frontispiece from *Mitchell's System of Modern Geography* (Philadelphia: E. H. Butler, 1866).

Chinese are remarkable for their order, industry, and regularity; but their treatment of females, their idolatry, and their general disregard of truth, lower them in the scale of nations and rank them below every Christian community." Likewise, the Persians, though "a handsome people . . . quick, lively, and versatile," were nonetheless "insincere and immoral," while the Hindustanis (East Indians) "are indolent and spiritless, have no patriotism, and are said to be nearly destitute of moral honesty." The civilized nations were acquainted with the arts and sciences and drew subsistence from agriculture, manufacturing, and commerce, yet these nations—including Spain, Portugal, Greece, and Mexico—were rife with class divisions. This conditional evaluation led easily to the final stage of enlightened societies of northern Europe and America, noted for their intelligence and mastery of enterprise, industry, art, and science. Benevolence was a striking feature in the national character of the British, while the French were "intelligent, brave, gay, and very military."[10]

Just as curious was the relationship maintained in the illustration *between* the stages of society. The savage and barbarous peoples were closely arranged, and the physical proximity between them implied easy and fluid exchange. At the bottom of the illustration stood the civilized, half-civilized, and enlightened societies, where a permeable division also suggested movement and mutual identification. Yet between the top and bottom very little evolution, ascension, or movement was apparent. Each stage was separated from the others and relationships only seemed possible between the savage and barbarous or between the half-civilized, civilized, and enlightened.[11]

The most powerful element of the environment in nineteenth-century texts, implicit in this hierarchy of nations, was climate. Mitchell's world, for example, was divided into climatic zones that corresponded to his stages of civilization. In the Torrid Zone, a 3,243-mile-wide "girdle" of land located between the Tropic of Cancer and the Tropic of Capricorn, the inhabitants were far less productive and industrious and—because of the extreme temperatures—more prone to suffer from disease and natural disasters. This hostile geography produced dark-skinned peoples who were "indolent and effeminate in their habits . . . seldom distinguished for industry, enterprise, or learning."[12] Alongside this description was a small drawing showing fierce animals in the foreground roaming through the tall grasses of the jungle. What initially appears to be another animal in the background is, upon closer inspection, a native of the Torrid Zone clad in a loincloth and scaling a tree. Mitchell contrasted this with the Temperate Zones, where the natural environment encouraged progress and vigor and contributed to the rapid advance and dominance of these cultures. What two recent scholars have observed of nineteenth-century characterizations of Europe generally is equally true of Mitchell's text: the Temperate Zones were framed as arenas of *possibility*, unlike the Torrid Zones.[13] The intensity of geographical determinism often varied according to geographical location, which meant that Europe and America were exempted from the "strict rule of nature" and thus hosted a wider range of human activity. By contrast, in Mitchell's text the Torrid Zones were directly conditioned and controlled by their environment, creating a kind of racial and climatic museum with a future that was, to a great extent, predetermined. These references to environmental determinism would become far more sophisticated, though equally racial, at the turn of the century.

Like the contemporary atlases, these texts generally *described* races and cultures rather than *explaining* them, and made few explicit connections between physical environment and material progress. Textbook authors frequently appealed to vaguely theological explanations for the status of human hierarchies but rarely fleshed out this relationship between the divine and the natural. More important than discovering this connection between individuals and their environment was simply giving order to each by classifying the human and physical worlds through racial typologies, climatic divisions, and the like. The very form of Mitchell's geography reflected this approach. The text was composed entirely of short questions and answers numbered and ordered so that teachers could easily drill their students on the stages of society, whether each race was located in the Temperate or the Torrid Zone, and which nations were the most civilized. Students were not asked to explain the origin and development of these "facts." In discussing the source of racial and cultural difference, Mitchell mentioned the influence of the natural world, but stopped short of linking physical conditions and human progress by concluding that there was much we simply did not know.

Thus Mitchell's world was relatively immutable. Though he lauded the "progress" and improvement of the United States, for example, he considered the possibility of other nations rising in the hierarchy of nations remote. Mitchell's static view of the world extended to his characterization of imperialism as simply a series of boundary changes rather than a historical, human process, and one that bore no resemblance to the American movement westward. Instead, this world differed in every regard from America, which was divinely separated by seas to the south, west, and east. Within this hierarchy of nations the United States represented a break from the tired decadence of Europe but also a civilization vastly superior to that of the native inhabitants. The world abroad was a spectacle to gaze upon from a distance, immediate only insofar as students were encouraged to identify their European roots.

Underlying nearly all of Mitchell's geographical explanations was a natural theology, wherein everything from the earth's rotation to the political organization of the world was attributed to divine creation. In fact, Christianity was just about the *only* dynamic force in Mitchell's geographies, spread through imperial contact. "The Christian nations are much superior in knowledge and power to all others, and, through the increase of their colonies, the influence of the press, and the exertions of missionaries, will no doubt, in the course of a few generations, spread their religion over the greater part of the earth." [14] Aside from the growth of Christianity Mitchell's phenomenally popular geography was generally ahistorical, aspatial, and one-dimensional. It was typical of its time.

Another highly regarded text of the period was Arnold Guyot's *Elementary Geography for Primary Classes*. This "home geography" began in the student's own back yard and then transferred these principles and relationships—through distances, directions, and maps—to other regions. Guyot opened his text with simple geographical terms and ideas and then moved outward, first to the United States and then abroad. Yet like Mitchell, Guyot organized information around the recitation of facts and the classification of races, nations, and continents. In the process, Guyot's own hierarchy subtly surfaced. In characterizing Europe he paid heavy attention to political organization and exports, while in Africa description centered almost entirely around wildlife. Like most other nineteenth-century texts, Guyot's geography was a means of observing God's handiwork. Furthermore, as Martin Lewis and Kären Wigen write, "Guyot saw the hand of Providence in the assemblage of the continents as well as in their individual outlines and physiographic structures." This might also explain the nineteenth-century tendency to emphasize the sheer grandeur of the natural world, manifest in the longest rivers, the highest peaks, and the widest oceans. Geography was a way to discover the natural order manifest in the physical world. [15]

What Guyot implicitly suggested—the deterministic importance of continents and their relationship to progress—was made explicit in Jane Andrews's *Geographical Plays*. A primary geography reader of 1880, *Geographical Plays* reified the importance of continental divisions through the relative emphasis placed on different regions. While Asia and Europe were both given individual stories, the author combined the treatments of Africa and South America because "if I gave a whole play to each of these continents, I should thus put them on a par with Europe, Asia, and North America, whereas the very fact that it takes two of them to make the subject of one play helps the child to a proper appreciation of their relative importance." In their combined play, "Africa" asked "South America" why they were regarded by others as "continents of the future." South America remarked that "people of to-day count on our present imperfections rather than on our past splendors, and say we have yet to live out a history, and with renown in the world." Africa readily agreed: "Why should we live in the past, when the future is before us, and we have resources to develop that will make it great?" While Europe was a theater of economic and political power, South America and Africa were continents of natural beauty, zoological diversity, and meteorological curiosity, continents that had "yet to live out a history." Andrews imagined a metageography where Africa and South America had essential, continent-wide qualities while Europe was a continent of diversity and complexity.

As in contemporary atlases, here too we find implicit claims about the determinative power of continents. The lack of progress in South America and Africa qualified both for European intervention, though the author hoped for a different—more American—model of governance, one that contained an implicit criticism of European imperialism:

> Africa: "Now, what will be the future of Africa? Will its old Egyptian greatness ever return? Will it become a great English colony in the south, or a French one in the north? Or will this little seed of freedom in Liberia grow, and spread, and make in the future a grand repubic?"
> South America: "A republic, I hope; for then I shall want to clasp hands with you across the sea."

Andrews drew a world of hierarchy, but one not without a future of change, a theme of fluidity that would become more prominent after the turn of the century.[16]

In its characterization of Africa, this text was more typical, echoing the description found in another popular reader from 1887:

> [T]here are mountains and lakes, and a great many tribes of black people: there are also elephants, lions, and a great many other fierce beasts. There are enormous birds also, and among them ostriches; but the truth is, that, besides knowing that all these are to be found, we do not know what is in the centre of Africa as we do of other countries. The heat, the unhealthiness, the jungles, and the fierceness of the people there, have kept out travellers. It is only very lately that we have learned anything about the country.[17]

Little differentiated this from the rote narrative of Mitchell's text, despite the author's claim of rejecting "the aimless and weary wandering" of most geography textbooks. William Swinton also claimed to avoid the treatment of geography as a morass of facts by incorporating the "heretofore neglected" subjects of industry and commerce. His popular geography of the 1870s would surpass past texts that focused on "jejune descriptions of the Irishman, the Italian, and Esquimaux, and the Chinese." Yet, like those he repudiated, Swinton also organized the world according to "stages of society" and levels of civilization that were linked closely to race.[18]

The character of these textbooks—lacking a clear analytical approach despite their implicit ideology—reflected the diversity of their authors, and of American education generally in the nineteenth century. Swinton had been a Civil War correspondent for the *New York Times*, and Mitchell was a publisher and author of tourist guides as well as schoolbooks. Samuel G. Goodrich, author of the popular "Peter Parley" series, had been a diplomat, a politician, and a publisher. Jedediah Morse, whose *Geography Made Easy* rivaled Mitchell's texts from earlier in the century for popularity, was a clergyman.[19] As the century closed, textbook authors became part of an increasingly professionalized culture, and after the turn of the century nearly all would be trained in their subjects at the university or teachers' college level. One early author described geography as "a way of travelling in our minds."[20] By the 1890s this approach to the subject would be challenged. At the turn of the century, school geography—like academic geography—underwent a transition from an idiosyncratic body of information to a systematic science. In both cases the goal was to emancipate the discipline from its thoroughly popular and amateur reputation as a subject concerned with names, places, and facts. This emancipation was a partial success.

The Reform Impulse in American Geography, 1893–1916

By the late nineteenth century enough high schools had appeared to generate concern about the character of education and the disparity of curriculum standards around the country. Secondary schools were increasingly responsible for educating both vocational and college-bound students, which raised the question of whether education would be primarily a tool of mental discipline or an instrument of assimilation, socialization, and practical training. This struggle reflected the contrary goals of those fighting for reform: labor leaders, progressives, and employers needed graduates who could understand their place in society, while university professors and other advocates of a tougher curriculum wanted students whose training and outlook would reinforce their own—young and as yet uncertain—social identity.

One of the primary means the latter group had of shaping the secondary school curriculum was the National Education Association. Initially made up of university professors, the Association aimed to improve high schools by establishing more rigorous and systematic standards for college-bound students. Though founded in 1857, the NEA remained insignificant until the membership of men such as William Torrey Harris, Charles Eliot, and Daniel Coit Gilman elevated its prestige in the 1880s. One of the organization's first and most significant efforts—undertaken in 1893—was designed to redirect the goals of secondary education. Led by Harvard's President Eliot and made up of college presidents, professors, and headmasters, the Committee on Secondary School Studies—also known as the "Committee of Ten"—wrote a report full of contradictions that reflected its inability to resolve the competing goals of secondary education. At the outset of the report the committee proclaimed its concern not just for those who planned to attend college but also for the growing numbers who would end their education with high school. Education in a democracy, it argued, would not divide society by separating college-bound and terminal students; instead all would take the same course of study. Having established this, however, the reformers intensely criticized the American high school for its lack of intellectual rigor, in both curriculum and instruction. In its place, the committee recommended an intensified program of analytical skills learned through history and science.[21]

The committee proclaimed the reorganization of geography to be the most far-reaching of all its plans, a reflection of its prominence in the curriculum. Though criticizing the weaknesses of geography teaching, it still argued—as had Halford Mackinder in Britain a few years earlier—that the subject's breadth had the unique potential to impart a general and practical knowledge of science. Encompassing botany, geology, zoology, astronomy, and meteorology, as well as elements of commerce, government, and ethnology, geography ought to be a comprehensive study that embraced all these disciplines when relevant.[22] Yet while endorsing geography's breadth, the committee was also under pressure to modernize the subject by offering students more than the litany of races, place names, exports, and census figures found in existing texts.

University geographers—well represented on the Committee of Ten under the leadership of William Morris Davis—were themselves building a scientific discipline, and their reforms reflect professional concerns. Some, such as Davis, hoped to isolate geography within the curriculum, reasoning that it would be best served by removing any trace of zoology or botany, restricting the subject to the evolving character of volcanoes and shorelines and the relationship between climate and land. Others disagreed, worried that an independent subject of geography could not survive in the schools. Among them was Ralph Tarr, one of Davis's former students who suggested that geography be correlated and integrated with geology, a school science with a more secure footing.[23] In their quest to strengthen the place of science in education and to secure their own intellectual reputation, the Committee of Ten endorsed the study of physical processes of the earth as the basis for both general science and geography. As argued in chapter 4, physiography marked a move away from earlier styles of geography that simply described the physical features of the land, toward a more scientific, analytic study of landforms. The subject would no longer lend itself to static classifications of the natural and cultural world, the reformers wrote, but instead would broaden the student's horizon by emphasizing landform evolution. Ultimately, these reformers endorsed both the independence of physical geography and its centrality to general science courses, reflecting in this duality a dilemma implicit in school and university geography. While embraced for its breadth, it was precisely this breadth that would prove to be geography's basic weakness.

A few years later the NEA again met to reconsider the role of geography in the secondary schools. In this report, the earlier focus on physical geography as landform evolution was reaffirmed, but was paradoxically both broadened and narrowed in order to strengthen its place in the curriculum. The NEA broadly defined physical geography as the natural environment of *human* life but simultaneously decided to exclude extraneous subjects that had come to be associated with it as a general science, such as astronomy, physics, geology, botany, and zoology.[24] A growing publishing industry was quick to incorporate both these recommendations and frequently hired NEA members to write the new textbooks. Among the first of these reformed geography texts was Alexis Frye's *A Complete Geography*, published two years after the Committee of Ten's findings. Frye made physical geography the center of a dynamic, evolving world where plateaus, valleys, mountains, and plains were created over time. This was a clear break from earlier notions of geography as a descriptive recitation of place names and geological categories. The text included some of the first relief maps made for students, and its solid physical geographic content made it popular in the schools for nearly twenty-five years.[25]

In the same year, Ralph Tarr introduced his *Elementary Physical Geography*, which also framed geography as a thoroughly physical science. Tarr developed these physiographic themes even further in his *Complete Geog-*

raphy, coauthored in 1904 with Frank McMurry. Here they emphasized principles of physical geography, but also began to focus on humans as more significant elements of the environment—not just coexisting with the land but products of it. This basic framework had been a crucial departure for university geographers, especially those training under Davis. Tarr, having been a student of Davis, was himself moving toward this principle of ontography and found little difficulty incorporating this vision into his textbooks. In fact, Tarr and McMurry's was probably the first textbook to initiate the twentieth-century conception of geography as a study of humans in a physical realm.

To some authors, the movement toward inclusion of man as a product of the physical environment had come even before the NEA's endorsements in 1898. Redway and Hinman's *Natural Geography* conceived of the world as a series of cyclical processes by introducing the student to concepts such as rising and sinking coasts, valley erosions, and shore forms. But the authors organized the text around the physical world as it related to *humans*, particularly their "history, customs, industries, and commercial interrelations as determined or modified by the inorganic forces of nature." Richard Elwood Dodge's *Reader in Physical Geography for Beginners* also framed physiography as it determined human behavior. Davis himself introduced a textbook in 1898 that defined geography as worthy insofar as it could be causally related to human life. The relationship between man and land, which had been so central to university geographers in their quest to professionalize, had begun to shape school geography.[26]

By 1898, a survey of high schools in Pennsylvania found that ninety percent required students to study physical geography. The NEA's recommendations had reflected general trends, placing physical geography on a par with literature, rhetoric, civics, mathematics, and Latin. But no sooner had the burgeoning textbook industry responded to these recommendations than the NEA began to express doubts.[27] The student body was becoming ever larger and more diverse, with a correspondingly smaller percentage attending secondary school as a path to higher education. In an attempt to address this new population's needs, to make education more socially efficient and to adjust its content to the nation's position abroad, the NEA again adjusted the curriculum.[28] This "second wave" of reform had limited consequences, in large part because of its ambiguous goals, and also because later reforms of the 1910s would overshadow any previous efforts. Still, its articulation of geography as a commercial subject proved the most resilient form of the subject in the modern American school.

The university-based reformers of the 1890s were concerned as much with the survival of their own disciplines as with the condition of public schools, and thus worked hard to influence the curriculum. But the subsequent growth of American education was accompanied by a more professionalized educational apparatus: pedagogy emerged as a professional discipline housed in a growing number of normal schools across the country, while increasing school enrollments gave administrators and teachers more authority and credibility within the NEA. This broadened membership of the NEA meant a new role for teachers who had more sympathy for the place of vocational education in a curriculum dominated by the college preparatory course. The rising standards embodied in the NEA's recommendations of the 1890s came just when schools were absorbing more students, fueling the need for vocational education. While the reforms of the 1890s had reconfigured the curriculum to include foreign languages, physical science, and composition, by the early twentieth century the NEA was endorsing courses in typewriting, bookkeeping, stenography, commercial law, domestic science, industrial arts, and manual training.[29]

Together, these trends—the diversification of the student body, the changing leadership of the NEA, and the rising influence of industrial interests—restructured the curriculum to provide an education that was, above all else, *relevant* to the student's future. This trend boded ill for physical geography. In 1909 the NEA again reviewed secondary schooling, appointing a separate committee to evaluate the worth of physical geography. That the committee was chaired by a normal school teacher rather than a professor or college president reflected the NEA's new leadership and direction. James Chamberlain had long criticized the existing structure of school geography, arguing that education ought to prepare students for life, and that geography ought to capitalize on the student's "natural" interest in human production and industry. Like other reformers, Chamberlain found physical geography—even when constituted as the environment of human life—narrowly construed, irrelevant, and dry. He insisted instead on an understanding of the subject that gave more weight to human interaction and social concerns. The relevance of geography lay in the earth's rotation, zones of climate and atmosphere, natural resources, the work of the government, and geography of the "most important" countries and populations.[30]

Constituted in this way, geography was not a science, but rather a social study, and the committee's legacy was to move geography in this direction by using physical geography to illuminate human behavior.[31] Academic geographers—at perhaps the height of their commitment to schooling—reinforced this turn by reconstituting high school geography around the ontographic relationship, which included both physical geography and the human response to this environment.[32] Ontography moved geography toward a general concern with the causal relation between humans and their earth, and then to an even more specific form of this relationship, one that centered on the human capacity to identify, extract, and manipulate natural resources. In the process, the NEA had weakened the place of geography as a natural science and strengthened its place among the commercial and social sciences.[33]

THE RESPONSIBILITY OF RESOURCES: GEOGRAPHY TEXTS BEFORE THE GREAT WAR

Geography textbook authors were highly sensitive to the new directions of curriculum reform and easily integrated the commercial focus into their materials. But the focus on a more utilitarian brand of education was furthered by the national climate of economic expansion as well. While the 1880s and 1890s perhaps represent the height of the European scramble for colonies, in the United States the question of overseas expansion remained undecided. But as the depression of the 1890s demonstrated the pains of overproduction, foreign markets looked ever more attractive. The sinking of the *Maine* allowed internationalists like Josiah Strong and Henry Cabot Lodge to seize the day and capitalize on public ire toward Spain's presence in the Caribbean and the Pacific. By 1900 active interests had been fueled in Latin America, China, and the Pacific.

The relationship between American foreign policy and the nation's textbooks may appear to be a truism: because imperialism and expansionism so dominated the understanding of the world—and because the United States was now actively participating—nearly every geography textbook would confront the issue at the turn of the century. But this relationship is complicated when we consider that history textbooks followed a different pattern. According to Frances FitzGerald, after 1900 history textbook authors frequently challenged the growth of American power abroad. Mugwumps, for example, portrayed the War of 1812 as unnecessary, an outgrowth of British and American shortsightedness, and similar criticisms were made of the Mexican War and the territorial acquisitions at the end of the century. Though FitzGerald concedes that twentieth-century history textbooks rarely made radical breaks from the general thrust of American foreign policy, she also points out that none was unconditionally supportive of either military or economic expansion abroad.[34] This contrasts dramatically with geography texts, which were forthright about American interests in the world after the turn of the century and rarely challenged these ventures. Though occasionally fears were expressed about the dangers of overseas expansion, the twin goals of internationalism—expanded trade and native uplift—won easy acceptance among geography textbook authors, manifest in their tendency to organize and evaluate the world according to commercial production. In the process, the character of the world began to move from a curious spectacle to an arena of commercial opportunity. Here, as in the case of the National Geographic Society and of the cartographic companies, geography was used to articulate the designs of the state.

The transformation of foreign interests at century's end, together with curriculum reforms, recreated geography texts in three ways. First, the textbooks reorganized the world around physical environment *as it related to trade*. Second, this focus led to texts that placed humans at the center through their ability to extract, manipulate, and export natural resources. Third, this use of resources became a measure of progress: authors judged states, cultures, and races worthy to the extent that they participated in trade. Thus by the early twentieth century this ability of humans to progress within their environment—to become *civilized*—became a function of commercial strength. Similarly, descriptions of progress began to hinge on the *conquest*, rather than simply the presence or absence, of resources.

References to commercial interaction in geography textbooks can be found as early as the Civil War, but only in the twentieth century was commerce integrated into the texts through studies of the earth as well as of human behavior. From the Spanish-American War to World War I these commercial geographies dominated the geography textbook market. Thus, even before the NEA and the AAG endorsed commercial geography, authors began to constitute geography as the story of economic life.[35] This emphasis on natural resources and commerce is the key to geography textbooks of this era because it not only linked humans with their environment but created a relationship that placed man at the center of geography rather than at the periphery, where he had stood under physiography and physical geography. The "humanization" of the curriculum, alongside the nation's expansionist drive, allowed commercial geography (and then political geography) to flourish in place of physical geography. Though a profile of the climate and export potential of the Philippines would be of no more immediate use to high school students in 1910 than one of their own region's economy and climate, the former affirmed the nation's economic imperatives. In this and other ways, commercial geography was suited to the demands of "practical" education.

On the most basic level these commercial geographies began to forge links and establish causal relationships rather than simply describe the state of the physical and cultural environment. Revising their geography text in

1904, Tarr and McMurry argued that they departed from the "beaten track" by explaining the physical world as it related to human needs and uses. They fused the physical landscape with a commercial focus by giving reasons for the geography of existing trade routes, the strategy of locating and constructing ports, and the like. In *Comparative Geography of the Continents*, introduced for elementary grades in 1904, Richard Elwood Dodge abandoned the old custom of teaching climatic "zones" with fixed boundaries of latitude, and instead began to focus on the interdependence of the continents through trade. In one of Dodge's later texts, the commercial profile even preceded the physical description of the land itself. Though titled "World Relations and the Continents," it was for the most part a series devoted to trade, with less information on government and culture. In 1912, Charles Redway Dryer's *High School Geography* organized the world into different human economies: "plucking and hoe culture, garden and horticulture, field culture, plantation culture, herding, hunting, fishing, lumbering, and mining." Put plainly, "man's desire to consume and his ability to produce have increased together with the progress of civilization."[36] Albert Perry Brigham, one of the country's foremost geographers at the turn of the century — trained under Davis and an author of the NEA's 1899 report on geography — illustrated the promise of trade by simply organizing his 1911 *Commercial Geography* around wheat, cotton, cattle, iron, and coal.

It was a short leap from granting commerce power as a means of progress to understanding it as a moral indicator. Brigham closed this text by arguing that industry and trade could modify "the habits, judgments, and policies of men and nations." Aside from bringing order and security to unstable regions, commercial commitments also required nations to build their financial integrity, invest in education, efficiently use their human resources, prevent internal and international hostilities, and develop sympathy with others. As Brigham wrote, "oppression in Armenia, or cruelty to natives in the Kongo [*sic*], arouses the feeling and elicits the protest of the world, and thus develops the common feeling of the human race in a degree unknown before the days of modern commerce." The purpose of his text was not simply to profile the balance of international trade, but "to present industry and commerce as organic, evolutionary, and world-embracing, responding to natural conditions and to the spirit of discovery and invention, and closely interwoven with the higher life of man."[37] In this vision, commerce *was* the nation. Despite his recognition of a clear racial hierarchy, Brigham envisioned a world where commercial strength could be harnessed for the good of all, relegating isolationism to the dustbin of history.[38]

This promise of trade is visually illustrated in Brigham's texts. *Com-*

Figures 5.2 and 5.3   These two full-page color images, celebrating geography as
the power of human innovation and trade, introduced pupils to one of the early
twentieth century's most popular geography textbooks: Albert Perry Brigham and
Charles T. McFarlane's *Essentials of Geography: First and Second Book* (New York:
American Book Company, 1916).

mercial Geography (1911) opened with a full-page color map of the Panama
Canal Zone, a symbol of the power of America—somewhat unique among
nations—to manipulate the natural environment, and a reflection of contemporary commercial confidence. In Brigham's later, two-volume *Essentials of Geography* (1916), the student finds more visual celebrations of commerce (figs. 5.2 and 5.3). Book One is introduced by a paean to engineering,
a landscape transformed—though significantly not disrupted—by technology. Peacefully integrated into the background is a train, while in the
foreground tiny figures observe a well-ordered dam that has harnessed a
river, creating hydroelectric power and regulating the irregularities of nature. Similarly, the frontispiece to Book Two is a majestic freighter docked
in a bustling harbor, an homage not to an architectural monument or a
natural wonder, as was common in early geography texts, but to a massive
cargo-carrying ship and its contribution to civilization. In both cases—
and in the map of the canal zone—the essence of geography is the human
ability to transform the natural world rather than the nineteenth-century

emphasis on racial difference, environmental determinism, and the physical environment itself. School maps also began to reflect this interest. By
1912, Rand McNally featured in its catalog of school products commercial
maps that highlighted the international distribution of resources and celebrated the nation's territorial growth.[39]

Frank Carpenter's contemporary geography readers also expressed passion for the ameliorative effects of trade, judging countries and colonies
worthy to the extent that they possessed commerce. In *Australia, Our Colonies, and Other Islands of the Sea*, Carpenter included first colonies and islands that had made a significant contribution to trade, while those without extensive commercial ties were given only passing reference. Carpenter
gave special attention to the "new world" of the West Indies and the
Pacific, particularly the islands of "our brown-skinned cousins"—Samoa,
Hawaii, the Philippines, Puerto Rico, the West Indies, and Cuba. In fact,
nearly forty pages were devoted to the Philippines, most profiling the population's habits and physical characteristics. Great effort was made to char-
acterize the Filipinos positively; they were "by no means bad looking . . .
[t]heir lips are not thick, and their noses are as straight as our own. They
look clean, and we learn that most of them take a bath every day." Furthermore, they had the intellectual potential for American citizenship,
"being naturally intelligent and anxious to learn." Despite this, Carpenter
could not help but convey his concern that people so different from him
and his fellow Americans—culturally, physically, and historically—were
on their way to American citizenship. In his view there was both fear and
hope for the future of the Philippines in an American empire.[40] Significantly, Cuba and the Philippines were even grouped together in the same
volume, along with Australia and other islands. It was not a geographical
grouping, but rather a series of colonial relations that bound these scattered islands together. Carpenter marveled at the fact that even in the
South Pacific he could still be standing on European soil. It was to him a
great achievement that, under imperialism, all regions were accounted for
by their place in a larger order.[41]

Yet the texts, not surprisingly, took care to distinguish American from
European intervention. The American "liberation" of Cuba and the Philippines from Spanish corruption was repeatedly raised in order to differentiate American from Spanish imperial interests. And while American expansionism was touted as more humane, justified in terms of American
responsibility and benevolence, it was even more commonly framed as simply a question of natural resource use. Tea, rice, and spices from China,
linen and wool from Europe, metals from Australia and southeast Asia—

these were the reasons for teaching geography, not idle curiosity about "strange and distant lands." As Tarr and McMurry argued, "our study of geography is chiefly concerned with Christian countries; for there we find the most varied and extensive uses of the earth in the service of man."[42]

This ideal of resource development as progress also shaped school maps prior to World War I. Rand McNally created maps of each continent specifically for educational use, and the company's marketing of these materials reveals its normative ideas of geography. The company's map of the Asian "continent"—which was separate from Europe—was described in the following terms:

> The partition of a large portion of the area among a few great powers is . . . graphically portrayed. The vast possessions of Russia, China, and Great Britain are seen in striking contrast to those of the few remaining native states. Asia's commercial awakening is plainly evidenced in the railway lines here shown.

The imperial interests in Asia could be easily taught to schoolchildren on such a map, in very concrete and identifiable terms. The word "awakening" is also suggestive here, and reflects the assumption that commercial development was the natural, inevitable course of history. That this awakening was evidenced by railway lines is also telling; railway lines, as argued in chapter 1, were historically central to American cartography, at times even more important than topographic features. On Rand McNally's map of Asia the presence of railway lines represents tangible evidence of development; put another way, the more civilized a region became—with towns, railways, and industry—the more it could be "mapped." According to this rubric, the continent of Africa was progressing in the early twentieth century, in both real and cartographic terms. As the catalog's description of the map of Africa read:

> The wonderful progress made during the last few years in opening up the interior of this great continent has so altered its appearance as to make this revised edition practically a new map.
>
> Before the coming of Livingstone the center of the continent, where rise the Nile, the Congo, and the Zambezi, could be covered, like some curious ancient maps, only with "elephants instead of towns." Now this region exhibits towns, kingdoms, tribes, vast lakes, rivers, and valleys, all accurately located according to recent surveys, sure evidence of a steady progress in African civilization.[43]

The mention of "elephants instead of towns" refers to Jonathan Swift's much-quoted verse:

> So Geographers, in Afric-maps,
> With savage-pictures fill their gaps;
> And o'er unhabitable downs
> Place elephants for want of towns.

The practice of filling blank spots on the map with animals and other descriptive text had begun to erode by the late eighteenth century, as geographers increasingly preferred the honest and scientific absence of detail to fanciful filler. So while parts of the African interior had once been covered only with "elephants for want of towns," by the nineteenth century the region was more commonly left blank.[44] Later, in the aftermath of Livingstone's exploration and the "opening" of Africa, the region began to exhibit "towns, kingdoms, tribes, vast lakes, rivers, and valleys." Parallel to the "awakening" of Asia through the mapping of railways, the newly discovered features of its landscape gave Africa cartographic legitimacy and demystified its identity. Yet in the African case these were not newly created feats of technology. Instead, Rand McNally was making real—in cartographic terms—what had already existed on the landscape: towns, tribes, lakes, rivers. Rand McNally's descriptions suggest that cartographic representations of the land had in some manner come to be more significant than the landscape itself.

Just as Africa, China, and South America were frequently positioned as the most backward and naturalized of geographical spaces, other areas were stereotyped in different ways. Yet after the turn of the century these assumptions occasionally ran counter to established racial wisdom, particularly in the cases of Japan and Spain. Carpenter regarded Japan as civilized because of its strict rules regarding addictive substances, its reputation as "a land of books and newspapers," and its adaptability.[45] The case of Spain in schoolbooks mimics the country's treatment in the *National Geographic* and in contemporary atlases: a country that had violated the mercantile code by failing to utilize the resources of its possessions. For instance, in describing the Philippines and Cuba, Carpenter repeatedly condemned the cruelty and inefficiency of Spanish administrators.[46] By contrast, the Japanese were endlessly invoked as exceptional among Asians for their mimicry and adoption of Western ideas and customs. Relative to Spanish indolence, Japanese ability to make so much with such limited resources made a strong case for the mutability of racial categories. No comparable differentiation between these two cultures can be found before the Spanish-

American and the Russo-Japanese Wars. In this sense there was a degree of dynamism and open-endedness to these explanations of the world: though environment and racial caste might *explain* a certain region's status, it did not chain these regions to their place. The twentieth-century texts describe a world that, if not egalitarian, was increasingly flexible.

The dynamics of this relationship were crucial. Humans were endowed with or deprived of an environment of either natural abundance or poverty, and over this they had no control. What they could monitor was the energy they brought to this environment, and this injected an element of fluidity into what in the nineteenth century had been predestined. In the 1870s and 1880s authors would have explained the resource wealth of the United States according to providential blessing—something bestowed according to the superiority of the Americans themselves. These texts rarely went beyond teleological typologies of race or culture to explain the hierarchy of civilization, but the emphasis on resources forced a modicum of dynamism into the way these texts constructed the world. The flip side to this flexibility, however, was that those who failed to maximize resources properly relinquished the right to govern their extraction and manipulation, and even to govern themselves. Imperial control was an issue of efficiency and resource management. Why any group failed to use its resources properly always involved a combination of environmental and biological—or racial—explanations. The imperative of finding a relationship between man and his environment led many toward suggesting a causal interaction between the two. As argued in chapter 3, after the turn of the century geographers relied on both longstanding theories of racial determinism and newer notions of environmental influence, such as climatology.

The vague nature of environmentalism, like Social Darwinism, made it especially resilient. In geography, it implied a relationship between physical and cultural surroundings on the one hand and human response on the other. Suggested in one form by Frederick Jackson Turner, and more strongly by Ellen Churchill Semple, for the most part it was left unarticulated in any systematic fashion, making it remarkably pliable. Applied to the study of commerce, environmental determinism became an evaluative framework that asked how humans managed to extract, manipulate, and export their natural resources given the climate, cultural customs, and other factors. From the turn of the century through the 1930s geography textbook authors clung to this framework but were not always loyal to its implications. Alongside it, vestiges of racial determinism coexisted in an uneasy, even contradictory relationship. While in theory these commercial geographies viewed the world through more egalitarian lenses, in fact they were just as likely to use the level of commercial interaction to reinforce longstanding notions of racial superiority and inferiority. In other words, if we claim that commercial geography brought a measure of dynamism into the study of international relations, we must also be careful to note the degree to which these often reinforced racial explanations.

In Latin America, for example, Tarr and McMurry explained that "the weather is too warm to produce energetic people." "So *little* energy is required to find sufficient food that the people do not *need* to exert themselves, and hence do not. . . . The people, therefore, lose the inclination to bestir themselves, or, in other words, become too lazy to improve their condition." But climate only partially explained the failure of Latin Americans to exploit their resources. Intermarriage between Spanish and Indians produced "half-breeds" that were "an ignorant class, far inferior to the Spaniards themselves, and so backward . . . that they still follow many of the customs of the Aztecs." It was this intermarriage, in fact, that accounted for the instability of South American governments. Conversely, avoiding miscegenation allowed the English to "act with more intelligence, speed, and force."[47] Both environment and heredity were invoked to explain the lack of progress in the southern regions, and this justified the dependent relationship of Cuba and Puerto Rico on the United States. Lamarckian theories of the interaction of environment and heredity sealed this coincidence of factors in the textbooks, fusing what might have been competing factors into a harmonious dynamic.

Striking such a delicate balance between environmental advantages and racial superiority was difficult. In a deft use of truism, Tarr and McMurry argued that occasionally environment predisposed populations to act in certain ways, while at other times racial characteristics survived and humans were able to conquer their environment. This explained the ability of South Americans to rebel against their oppressive Spanish governors, even though "the nature of the population was such that real republican government was impossible." Similarly, Carpenter argued that the

> great race of Europe and the great race of the world of to-day is the Caucasian. It is our race, the race which has done most of the work of the civilized world, and which promises to control the whole world in the future.

On the next page, however, Carpenter admitted that "they [the Europeans] have so many natural resources, and their situation is such, that they could hardly help reaching a high state of civilization and power." When describing Germany, Carpenter suggested that "it could hardly help but being the home of a great people," yet added that more than land was necessary for success on this scale, invoking the thrift and enterprise so characteristic of Germans.[48]

The challenges to racial determinism had grown steadily in the twentieth century, evident in the increasing opposition to Social Darwinism. And in most textbooks, references to racial influence were diluted by references to environment. Yet unmodified racialism continued to surface as late as the 1920s. In Dryer's *High School Geography* the author included a detailed description of the different races, their natural regions, and their cultural characteristics. Significantly, this was placed in the "physical geography" section of the text, effectively naturalizing what were often cultural distinctions. At times the environmentalism of the texts was implicit in their organizational structure. The *Rand McNally Elementary Geography*, in print from 1894 to 1907, introduced students to geography through a study of the physical elements—the size of the earth, land forms, ocean forms, topography, climatic zones, and zoology—and only then moved to place humanity within this physical context. As a result, the text implied that humans were an outgrowth of their physical surroundings. At the same time, however, the text functioned according to a strict, almost naturalized, racial hierarchy.

> Most of the civilized people of the world belong to the white race, though in some countries the people of that race are half-civilized. The savages belong to the red, brown, and black races. Most people of the yellow race are half-civilized, but you will read some day of the yellow people of Japan. They are the only great people of that race that has become civilized, and in recent years they have adopted many American customs and become very powerful.[49]

This kind of commentary in the early twentieth century reflects how little some aspects of geographical education had changed since the mid-nineteenth century.

This tenuous balance between environmental and racial explanations is captured perfectly in the following passage from Tarr and McMurry's text:

> As the environment of the desert has given rise to the nomad, and the ease of life in the tropical forest to the degenerate savage, so the environment in the United States has given rise to a race noted for its energy and enterprise. This race has been possible, however, largely by reason of the fact that it comes from a mixture of peoples already gifted. That resources alone will not make an energetic people and a great nation is well illustrated in China,

where nature favors, but racial characteristics and customs are
opposed to, development.[50]

Upon examination, the logic of these paradoxical explanations becomes
clear. To explain American success solely through environment would fail
to distinguish it from China, a nation also blessed with natural abundance.
But conversely, to invoke only the superior character of a distinct American people—which in 1900 was problematic given the diversity brought
by immigration—would fail to set them apart from their less successful
European counterparts. This delicate balance of environmental and racial
determinism was a way for American geographers to explain their own
condition as well as that of the people around them. Never content to explain the state of a people, region, or nation simply according to race, these
geographers were also products of their time, unable to ignore the power
of racial explanations altogether. Competing explanations of race and resources would last, in various guises and levels of sophistication, in geography texts well into the 1930s.

Another striking continuity in the history of geographical education
has been its emphasis on contrasts and difference. Ultimately, geography
made sense of the world for students insofar as it successfully compared the
foreign and the familiar. Even in the late eighteenth century, Jedediah
Morse described the new nation by contrasting the industrious and virtu-
ous New England Congregationalists with dissolute New Yorkers, uncul-
tured Baptists, and decadent slave holders. Later, in the post–Civil War
industrial climate, another textbook championed the rural life against
both European and American cities. And just after the turn of the century,
a third text used Henry Van Dyke's "America for Me" to introduce students
to the contours of world geography.

'Tis fine to see the Old World, and travel up and down
Among the famous palaces and cities of renown
To admire the crumbly castles and the statues of the kings,—
But now I think I've had enough of antiquated things.

So it's home again, and home again, America for me!
My heart is turning home again, and there I long to be,
In the land of youth and freedom beyond the ocean bars,
Where the air is full of sunlight and the flag is full of stars.

Oh, London is a man's town, there's power in the air;
And Paris is a woman's town, with flowers in her hair;
And it's sweet to dream in Venice, and it's great to study Rome;
But when it comes to living, there is no place like home.

I like the German fir-woods, in green battalions drilled;
I like the gardens of Versailles with flashing fountains filled;
But, oh, to take your hand, my dear, and ramble for a day
In the friendly western woodland where Nature has her way!

I know that Europe's wonderful, yet something seems to lack;
The Past is too much with her, and the people looking back.
But the glory of the Present is to make the Future free,—
We love our land for what she is and what she is to be.

Oh, it's home again, and home again, America for me!
I want a ship that's westward bound to plough the rolling sea,
To the blessed Land of Room Enough beyond the ocean bars,
Where the air is full of sunlight and the flag is full of stars.

Van Dyke's verse, like the other examples, describes place in terms of its opposites and contrasts. Given this characteristic, we might ask how geography clarifies the identity of the local in relation to the distant, and makes sense of the distant in relation to the local. As Edward Said observes, the "other" has meaning only in relation to the familiar. The European, for instance is understood insofar as it constrasts with the American. Similarly, the adventures in exotic reaches make sense only if we can safely return home.[51]

As the American school became a means of democratic assimilation, its work became increasingly contested. Opening geography to redefinition allowed it to be influenced by American expansionism, school policy, and academic geography. While early textbooks constituted the physical environment primarily through climate and topography, after the turn of the century this definition was broadened to include natural resources. Correspondingly, textbooks began to explain human progress and civilization not strictly as a function of climate and race, but in terms of climate and commerce. Geographers accommodated this framework by explaining commerce as a function of human will and natural abundance. By holding out the possibility of altering one's physical constitution through behavioral modification and environmental adjustment, geographers justified imperial intervention, in both the abstract and the concrete. Furthermore, by framing the world in terms of Lamarckian environmental-biological evolution, geography textbooks ensured the survival of these ideas long after they had been discredited by most academic geographers. After World War I, the texts turned toward explanations of human-environment interaction that were less deterministic, though no less subject to external influences.

PART 2

Geography for the American Century

6

School Geography in the Age

of Internationalism

1914–1950

Shortly after World War I, the *Atlantic Monthly* ran an article by Edward Yeomans excoriating the sorry state of American school geography. Voicing a refrain that would be echoed by later generations, Yeomans wrote:

> The geography teacher is a girl of twenty-five or so, who . . . chose geography because she might just as well teach that as anything, and she seemed particularly good at remembering the boundaries of things and the principal rivers. . . . The geography teacher has a map on the wall. When the map is there, the children are asked questions like this: "What are the main exports of the State of Massachusetts?" When the map is not there, the children are asked to bound the various states—to give the names of the capitals.[1]

Yeomans's sarcastic tone suggests that his complaints about poor teacher training and the rote character of school geography were relatively familiar to his readers. In fact, as argued in chapter 5, educators had for decades attempted to reform both the theory and the practice of school geography, usually with limited success. Furthermore, Yeomans's observations still resonate with us today, an indication that our own stereotypes about American geographical illiteracy have a long history. Even more interesting is the way the author continued his critique:

> It would be something . . . if you could get the geography of the Malay Archipelago, for instance, taught by some native friend of Mr. Conrad's; if you could get Sven Hedin or Ekai Kawagouchi to pick a man from Thibet to teach the children about the Himalayas.

But no—they must be taught by someone who prefers the security
of a flat to the rigors of climate on the open surface of the earth
under the windy sky.[2]

In fact, this critique was at odds with itself. Notwithstanding the poor
teacher training Yeomans found so frustrating, his own ideal had perhaps
become anachronistic. As chapter 5 argues, reformers had made every effort to transform geography from "aimless and weary wandering" into a systematic, rational body of knowledge. But in becoming a science of human
relationships, school geography would necessarily lose some of its adventurous appeal, which—ironically—became largely the province of the National Geographic Society by the 1920s.

As Yeomans's comments suggest, classroom realities often eluded the
energetic efforts of the National Education Association, and criticisms of
geography's rote character continued long after it had been reconceived as
a science of human relationships. Though textbooks made strides toward
geography as a dynamic explanation for human behavior—by incorporating environmental determinism and by treating geography as the study of
commercial interaction—older traditions persisted. Further, the reforms
wrought between 1893 and 1910 contained contradictions and ambiguities
about the purpose of geography in particular and the curriculum in general.
"Landmark" reports in 1893 and 1909 attempted to reconcile the needs
of college- and non-college-bound students without acknowledging fundamental differences between the two. These reforms, combined with the
character of American interests abroad and the declining influence of academic geographers, brought school geography from the arena of physical
science and the natural world into the human realm of the social studies.
By making geography a "human" science in the early twentieth century,
educational reformers had effectively opened the way for an entirely new
set of concerns to inform the subject, the first of which was commerce and
resource development. By the late 1930s, events abroad again reconstituted
the subject, pulling it away from its commercial focus toward a more geopolitical and expressly internationalist focus. Understood together, these developments over the first half of the twentieth century highlight the historically fluid nature of geography in American schools.

THE REORGANIZATION OF AMERICAN EDUCATION

The progressive concern for education prior to World War I sparked a
number of studies across the United States, many of which echo Yeomans's
lament about the state of school geography. After participating in John

Dewey's studies of schools in Gary, Indiana, Randolph Bourne complained that "the children read over the assignments chiefly with a view to finding the answers to the questions printed at the end of the section. . . . The teacher, with book in hand, puts seriatim the above mentioned questions, occasionally adding one or more on her own initiative." Despite the reformist fervor of the early twentieth century, the lesson Bourne witnessed seemed "little more than a sight reading exercise."[3] Stanford professor of education Ellwood P. Cubberley concurred. After studying the schools of Portland, Oregon, Cubberley found the curriculum caught between the nineteenth-century goal of education as a tool of mental discipline and the twentieth-century notion of the school as an instrument of social utility.[4] To improve the schools, he recommended an overhaul of the geography curriculum, since this subject had more potential to excite the student's interest than any other, though in practice it too often became "abstract and bookish in the extreme." Most teachers simply divided the textbook over the course of the school year, making few connections between book learning and the students' own environments. At examination time, Cubberley, like Yeomans and Bourne, found nine- and ten-year-olds asked to define "continent," name each one, and identify the races of their inhabitants. Little distinguished this style of geographic education from that which had reigned in the 1880s. In light of this, Cubberley advocated a curriculum — for both general and vocational study — that replaced the mechanistic, rote course with one oriented around the students' lives, one that emphasized the nations and regions with the closest economic ties to the United States.

These individual reformers, along with the NEA reforms prior to World War I, foreshadowed the evolution of schooling toward a social rather than an intellectual enterprise. In 1916 the NEA's Commission on the Reorganization of Secondary Education introduced the new social studies curriculum, which dismissed traditional goals such as mental discipline in order to champion newer ideals of assimilation and preparation of citizens for life in an industrial society. The Commission included twenty professors of history, nine superintendents and principals, three teachers, and two professors of education. The conspicuous absence of geographers was no accident, but rather a reflection of geographers' fears that the new social studies would undermine geography's independence in the curriculum. Denying the existence of the social studies, however, did not make them disappear, and the absence of geographers only made it easier to reform education with their subject in a subordinate position. Thus the Commission's recommendations—which eventually became a national model—reflected the dominance of historians. History would be taught from the seventh to the twelfth grades. In the seventh grade geography would be a half-year independent course or a supplement to history courses. In the eighth grade geography would also be taught incidental to history; and in the tenth, eleventh, and twelfth grades it would be taught as it related to European history, American history, and civics.[5]

The second of the NEA's major reports, released in 1918 as the culmination of a three-year effort, has come to be regarded as the essential statement of modern education. The "Cardinal Principles of Secondary Education" claimed that the composition of the student body had dramatically changed over the previous two decades, and though one third of all students who entered first grade reached high school, only one of nine graduated. Education had to address the needs of this population, and the report brims with Deweyan optimism about the capacity of education to develop the individual's "knowledge, interests, ideals, habits, and powers whereby he will find his place and use that place to shape both himself and society toward ever nobler ends." Thus it was unsurprising that the NEA went on to ground education in social rather than intellectual objectives: health, command of fundamental processes, worthy home-membership, vocation, citizenship, worthy use of leisure, and ethical character were the new cornerstones of the democratic school.[6] The "Cardinal Principles" were not the first instance of the NEA's endorsement of vocational education, as chapter 5 pointed out, but they unambiguously defined education as a foundation of democracy. Together with the report of 1916, the "Cardinal Principles" so emboldened reformers to restructure the curriculum that twenty-five years after the Committee of Ten the "official" goals of the American school had been completely reorganized.

Geographers worried about these developments even as they distanced themselves from curriculum reform. Many were concerned that participation in the NEA's work would only ensure geography's subordination to history and civics. Though some geographers organized to fight this trend — Richard Elwood Dodge warned geographers that "we cannot play ostriches any longer"—most were interested primarily in cultivating their academic and professional reputations at the university level.[7] Furthermore, many considered commercial geography particularly unpromising as a professional field despite — or perhaps because of — the fact that it was becoming the most resilient form of the subject in schools. In an effort to concentrate on research, the venerable American Geographical Society relinquished control of the only journal devoted to geographic education, the *Journal of Geography*, a move that further distanced professional geographers from schoolteachers. The wider that gap, the weaker geography became relative to other subjects. By comparison, professional historians were active in school reform, successfully dominating the new National Council on the

Social Studies and its journal, *History Teachers Magazine*.[8] Unlike historians, university geographers found little incentive to join the fight for their subject's independence even as they bemoaned its position within the social studies.

The First World War heightened the stakes in American education by demonstrating just how thoroughly the nation was involved in international trade. This realization spawned education reform, especially when rising school enrollments made a pragmatic curriculum less an ideal than a necessity.[9] The war was also evidence that Europe, long held up in textbooks as the embodiment of civilization, order, and prosperity, was simultaneously a place of unparalleled destruction. This grim realization brought to American attention not only the extent of international commercial and political interdependence, but also the nation's responsibility to lead this new world. Overall, the war turned the public's attention toward world trade, and toward the concept of world citizenship inspired by the Paris Peace Conference and the League of Nations. At Versailles, new nations had been created and empires dismantled. Together with advances in telegraphy, cable communication, rail, radio, and aviation, these upheavals sparked a reconsideration of space that often translated into warnings about American geographic illiteracy. Laments about the dangerously low level of general geographic knowledge proliferated after the war in both the professional and the popular literature. As one educator wrote, no longer could American "smug self-complacency" be indulged; Americans had no choice but to preserve the armistice by introducing "world citizenship" to the young.[10] Wallace Atwood, in his inaugural address as the president of Clark University, scorned American geographical ignorance as an intolerable situation given the nation's newly inherited international responsibilities. Others emphatically argued that the war—by literally recreating Europe— demanded an awareness of independence movements emerging from the old order. Geography was uniquely positioned, educators argued, to inculcate this new world order. Gaining a sense of how the world worked— primarily through trade—was touted in the interwar period as the most valuable contribution geography could make to education. Leonard Packard, in advocating geography as the study of commerce, framed trade itself as a source of diplomatic peace. What distinguished all these calls for revitalized school geography was their explicitly human cast: geography had become understood as altogether commercial, human, international, and comparative.[11]

These calls for a more cosmopolitan education won geography temporary attention in the curriculum. But no statements of urgency could match sustained involvement in the mechanics of reform, and as the social stud-

ies gained momentum after the war geographers became even more indif-
ferent to the fate of their subject outside the university. Harlan Barrows,
one of the nation's leading geographers and later president of the Association of American Geographers, criticized his colleagues for their detachment, while another even questioned their motives:

> I have the feeling that the geography group has not been very much
> alive to the situation and has offered little in the way of a con-
> structive program. Why is it? What do they expect is going to hap-
> pen? Are they counting on geography retaining its time-honored
> place in our curricula? Just what do they think entitles it to the
> time and energy it commands in various parts of the country?[12]

At the dawn of the century many school reformers had been drawn to the
idea of a general science course for high school freshmen, a niche that
physical geography often occupied. As seen in chapters 4 and 5, geography's character as the "mother of all sciences" had begun to be questioned,
by other scientists, reformers, and geographers themselves. These professionals especially doubted the relevance of this breadth in a new era of specialized knowledge. In light of this, William Morris Davis lobbied for a nar-
rower definition of physical geography in the schools, but geographers
active in curriculum reform still held out the hope that their subject — naturally broad and synthetic — might be made the basis of school science. For
many educators after the turn of the century, the idea of physical geography as the core of the new science curriculum was little more than a joke,
and overarching claims to be "the mother of all sciences" only hastened its
demise. G. Stanley Hall considered geography "the sickest of all sick topics of the curriculum"; with "all the unity of a sausage," it was called a subject for those "who crave to know something, but not too much, of everything."[13] Asserting breadth in the age of specialization clearly had its perils.
If geography was the mother of all sciences, it was vulnerable to matricide.

As traditional courses such as botany and physics became more popular alongside newer school sciences such as astronomy and geology, geography's dismemberment seemed not just logical but imminent. A decade after it was hailed as the most democratic of the sciences for its sheer breadth,
physical geography was superseded by more specialized subjects in the sciences and by more "useful" forms of geography in the social studies.[14]
Daniel C. Knowlton, professor of history and civics at Teacher's College
and one of the chief architects of the new social studies, castigated geographers for their constant bickering over the best form of the subject and
their unrealistic demands that it be considered both a social and a natural

science. For Knowlton and others, the assumption that geography would remain a stronghold of the curriculum without the active and unified involvement of its professional practitioners was absurd. If geographers could not assert a simple and feasible approach to their subject, Knowlton believed it would be done for them.

> When we are told that "there are no facts or phenomena which are the exclusive property of geography" and that "the subject has therefore a different basis from all other branches of learning except philosophy" we feel even more hopeless of a working definition which will prove intelligible to the rank and file of its exponents in the schools.[15]

Many geographers failed to realize the depth and breadth of support for the social studies, and continued to resist this format throughout the 1920s.

These developments—the relative indifference of geographers, the reforms of the NEA, and the urgency of World War I—had three effects on school geography. First, geography's human dimension was incorporated into the social studies as a supplement to history. In light of the past, the new relationship between geography and history was especially ironic. Closely correlated in the nineteenth century, the two subjects had been separated by the NEA's reports in the 1890s, only to be reunited again in the 1920s, with history at the fore. Second, physical geography lost its independence and its elements were soon absorbed into a general science course. Finally, the sole form of geography to survive as an independent course was commercial geography, which by the mid-1920s had in many instances incorporated general economic geography as well. In one survey from 1906 to 1911, physical geography was found in thirty-three of forty schools, and general science in only one. Less than a decade later, from 1915 to 1918, only twenty-one schools offered physical geography while nineteen had turned to general science courses. The proportion of these schools offering physical geography courses fell from 90 percent in the 1890s to just over 50 percent by the 1910s. In Cleveland, 525 high school students were enrolled in physical geography in 1900, but by 1909 the number had fallen to 292.[16]

As the reforms of the 1910s eroded the physical dimension of school geography, educators began to use the subject to explain the earth as the home of man and as an introduction to the human sciences. In both cases, geography was primarily economic and commercial, reinforcing a trend begun at the turn of the century. Commercial geography was bolstered by the acquisition of overseas colonies in the Spanish-American War and then by the graphic demonstration of economic interdependence in World War I. In a study of geographic education in New Jersey, 73 percent of high schools had not offered a commercial geography course in 1905. But by 1915, six years after the NEA's endorsement of the study, only 50 percent had not offered such a course, and by 1925 only 34 percent of those schools responding offered no commercial geography course. Conversely, while in 1905 21.5 percent of all high school students enrolled in physical geography, by 1922 the subject attracted only 4.3 percent, and by 1934 only 1.6 percent.[17]

NATIONS AS NEIGHBORS IN THE INTERWAR CLASSROOM

As these statistics suggest, the most salient feature of interwar geography textbooks was their treatment of the world as a commercial and economic unit, continuing and expanding a trend begun in the earliest years of the century. Almost entirely gone were references to physical geography except as it affected human life, and few could miss the trend toward social problems after the reforms of the 1910s. The texts were also resolutely internationalist, celebrating the role of the United States in the Paris Peace Conference, its possessions in the Pacific and the Caribbean, and its commercial and economic networks worldwide. Corresponding to this was a sharper focus on the relationship between humans and their environment through the mechanism of trade.

These profiles of international trade were often introduced to broaden the student's horizon and inculcate the sense of "world citizenship" so important to contemporary reformers, while at the same time endorsing the nation's widening economic interests. *The Working World* focused exclusively on commerce by organizing the world according to regions and commodities, as in this description of the United States:

> The Pacific Northwest is a region of thriving sea ports, magnificent forests, extensive forest industries, and of abundant water power. East of the Cascades lie the pasture and wheat lands of the Columbia Plateau. The Rocky Mountains come next, where marvelous scenic and mineral resources have given rise to national parks, resort towns, and mining camps. The section of the Great Plains crossed along this northern route is much like the Texas portion. Grazing is the principal land use in the western part, wheat growing in the eastern.[18]

This geography text concluded with a ranking of the great producing and manufacturing nations, and most all the maps focused on the location of industries, commodities, and settlements, and on population distribution. Packard and Sinnott's *Nations as Neighbors*—as the title itself suggests—reminded the reader that in the modern world of heightened proximity each nation had something valuable to contribute. The opening chapter, "How Nations Depend upon One Another," illustrated progressive advances in communications and aviation that, though sponsored by the United States, benefited all.[19] The Depression, in threatening many of these economic wonders, redoubled the prescriptive emphasis on healthy trade and commercial cooperation and raised concerns that students of the 1930s would be robbed of the chance to see "normal" trade conditions.[20]

This commercial model of geography was also manifest in the interwar classroom maps. Many labeled "political" maps did include national or imperial borders but in many other regards were shaped by commercial concerns. Significantly, Rand McNally produced trade maps of Mexico, the Caribbean, and the Gulf region but not physical maps (i.e., topographical and climatic) of these same areas, which suggested that the area was primarily relevant as an economic entity. Furthermore, the sensitivity to interdependence after the war was reflected in maps that placed the continents in a wider geographic context rather than continuing to map them as discrete entities. On the political map of the world for schools, North America was placed at the center, with trade routes that swept out to the east and west. And since Rand McNally declared that "it is no longer possible to study one country, or even one continent, without a view of it in its setting in relation to other countries and continents," the map was also available with North America positioned at the left and Eurasia left undivided at the right. By the same logic the "Goode Political Europe" map was arranged to emphasize the "dependence of Europe on the neighboring land masses," a premise fairly inconceivable in earlier decades.[21] In the 1910s, Rand McNally also marketed a series that mapped the distribution of world products, steamship routes, railroad lines, canals, and mine locations.[22] And in Rand McNally's Goode Map Series, introduced in the early 1920s, a new map of Australia and the Philippines was justified in light of the potential of Asian commerce. The map placed the Philippines at the center, "thereby giving us a clear picture of their position with reference to all countries in their immediate vicinity. . . . [while t]he great trade routes from the ports of the Orient to the rest of the world . . . compel attention to commercial activities by water in the Far East."[23]

This pervasive theme of commercial interdependence extended to the descriptions of imperialism itself in the narrative of textbooks. Rather than a strategic or political endeavor, imperial intervention continued to be approached primarily as commercial, benevolent, and justified by the in-ability of natives to make efficient use of the land. One textbook of the late 1920s referred to American territories entirely in terms of resources, while another asked students "why the British became interested in the lands of Asia, and why these lands are valuable parts of the British Empire." By teaching Asians better methods of farming, by helping them build roads and railroads, improve harbors, and develop foreign trade, the British had uplifted their Eastern neighbors.[24] Similarly, India was "A Country Won through Trade," and Europeans controlled most of Africa because

> most of the groups of native people are very backward, and they
> have never learned to make the best use of the lands in which they
> live. Therefore, when white men from Europe came to Africa and
> began to develop the farm lands and grazing lands and to open
> mines and build railroads, it was natural that they should become
> the rulers of the native peoples.[25]

Perhaps the most popular interwar text, Edward Van Dyke Robinson's *Commercial Geography*, opened with a proud profile of American expansion into the Pacific, from Seward's purchase of Alaska to the Spanish-American War, all of which aided the quest for Asian markets. As he wrote, "When Dewey's guns in Manila Bay sounded the death knell of Spanish dominion, they thus at the same time gave to the United States an 'empire on which the sun never sets,' and unbarred to American commerce the Gates of the Orient."[26] The nation had been expanding since Jefferson's presidency, Robinson argued, and the newest acquisitions were but small steps in the continuing growth of the empire.

Yet within the textbooks there were also indications that American confidence in imperialism had been shaken. By repeatedly emphasizing the stability of trade and the need for international cooperation to maintain peace, the texts silently invoked the recent European catastrophe. In fact, some texts began to face the question of colonial independence by directly addressing the legitimacy of foreign rule. One text made it clear that the United States and Britain did not "own" colonies but rather were trustees overseeing the efficient extraction and manipulation of resources. In profiling the Philippines, they concluded that while the stark climatic and racial differences between the islands and the United States provided a rationale for Philippine independence, the islands still demanded supervision. Such outright rationalizations rarely appeared in prewar texts.[27]

To explain why certain societies were more or less industrious, the textbooks continued to draw on both racial and environmental frameworks. In

Packard and Sinnott's *Nations as Neighbors,* for example, the authors referred to the "invigorating effect" frequent storms and cold waves had on northern Europeans, "making them feel strong and ready for much hard work."[28] Rand McNally's advertisement for a new series of geographical readers published in 1928 also suggests a continuing environmental determinism, especially in the description of the African text:

> Like the other books of this series, the purpose of *Africa: A Geography Reader* is to show the dependence of the peoples of the continent on the geographical conditions under which they live. Stress, therefore, is laid on the facts which explain why, in the water-poor regions of the Sahara, the people are few, fierce, and mostly nomadic; why level tracts of land with a good rainfall are occupied by tribes of industrious agriculturists, among the most efficient laborers in the world; why conditions of life on the Nile made the river valley the home of one of the most influential of early civilizations, and why, on the contrary, the overpowering forests crushed human initiative, kept its inhabitants backward, and in time dwarfed them into pygmy races.[29]

Using the phrase "peoples of the continent," the advertisement made Africa *the* primary unit of understanding, irrespective of the enormous differences in politics, climate, topography, and culture within that vast space. It also made the physical fact of Africa the primary influence over the population and culture. Significantly, this series of textbooks—which emphasized the physical environment in relation to human activities—covered Latin America, Asia, and Africa, but not North America or Europe. This underscored the decision to treat the first three as places of limits, naturalizing the continents through descriptions that hinged on the organic relationship between climate and human activity. By contrast, the latter two were continents where the environment was a complex of both human and physical dimensions, and were generally regarded as places of possibility. Robinson's *Commercial Geography* nicely captured this theme.

> Nations come and go like leaves on the forest trees; the earth remains. . . . It would seem, therefore, that no nation can have a perpetual title-deed to any part of the earth. It belongs to mankind. At all events, few will seriously maintain that the rest of the human race must forever be denied access to the riches of a land because the inhabitants, perhaps a few naked savages whom chance has placed there, will neither develop the resources of the country themselves nor suffer others so to do. . . . The European

> conquest and partition of Africa, though far from justifiable in
> many of its methods and incidents, is therefore on the whole a
> justifiable process, carrying light into dark places and placing
> immense natural resources at the disposal of mankind.[30]

Robinson both acknowledged the controversial character of imperialism and made clear that improper use of natural resources justified foreign intervention even when this undermined self-determination and sovereignty. His explanation of how different societies used their resources hinged upon race, physical environment, and cultural traditions.

Perhaps reflecting the slow intellectual growth of academic geography, textbook authors continued to refer to environmentalist explanations of human behavior well into the 1930s, though after World War I these explanations became somewhat more muted. One of the most common interwar textbooks, *The Working World*, explained the range of human productivity through a number of factors that included but did not privilege climate—such as access to capital, cultural standards of living, and the like. When describing biological elements of human geography—such as race—care was taken not to connect these to respective levels of productivity. Thus when speaking of the English, *The Working World* attributed their success not to providence or racial superiority but to geographical position, a forgiving climate, and resource wealth. In describing the desert's influence on nomadic peoples, J. Russell Smith suggested that the Arab was driven to plunder not because his race linked him to a debilitating climate, but because his general environment allowed few other options.

> The man who is starving has little thought of right or wrong. To
> have such thoughts would seem to him fatal. . . . In a word laziness
> according to our definition of the word, is no great disadvantage
> provided a man is able to summon up his powers in a crisis when
> the camels have strayed far away, when they have been driven off
> by raiders, or when the man himself goes on a foray. Hence the
> Arab is lazy as well as utterly disregardful of the commonest principles of honesty.[31]

In all these examples, geography had become predominantly a commercial narrative. J. Russell Smith's *Commerce and Industry* divided the nation according to industrial rather than regional or political units. In this instance, geography had become so thoroughly commercial that its most basic principle—spatial relations—was made secondary to trade networks.[32] Not until World War II, when American political stability was directly threatened, would geography be forced into a more explicitly spatial framework.

Aviation, War, and the Politicization of Space

In 1927 Charles Lindbergh stunned Americans with a transatlantic flight from New York to Paris, one of the many heroic feats of aviation undertaken in the late 1920s and 1930s that included Admiral Byrd's flight over the South Pole, Amelia Earhart's transatlantic journey, and the Graf Zeppelin's circumnavigation of the world. These expeditions captured the imagination of the young, reflected in the popularity of model airplane "clubs" nationwide such as the Junior Birdmen of America. The proliferation of radio also fueled this trend through programs such as the Jimmie Allen adventure series.[33] But the romantic, adventurous character of aviation began to change after the rise of military dictatorships and armed conflict in the 1930s. The Second World War underscored the strategic — rather than the adventurous — dimension of flight, manifest in the changing character of Jimmie Allen's adventures. Eventually flight became less a source of dreams than an instrument of national necessity.[34] In an effort to encourage American students to compete with their European counterparts, the Rockefellers set up the Air Youth of America in 1940. In 1946 an international congress on air-age education was held in New York, and airlines were publishing weekly school newspapers to promote the same.[35] Hastened by the exigencies of war, "aviation education" was widely embraced by both the Office of Education and the Civilian Aeronautics Administration, and thereby spread quickly via air-age textbooks that were designed to adjust students' sense of geography to fit new technological and political realities. This frenzied interest in the air age — together with the war — demanded a dramatic reorientation of space and distance that encouraged interest in geography during the 1940s. The discipline gained some independence from the social studies, continued to demonstrate its potential as a commercial subject, and moved toward a spatial focus manifest in the rise of political and air-age themes. Most importantly, the war infused geography with a sense of urgency and a spatial reorientation; what students encountered in mid-century textbooks was unlike anything that had come before.

Secondary-school geography still focused on commerce and economics in the early 1940s, but by the middle of the decade the number of courses devoted to geography as a political and spatial study had risen dramatically. During 1942 nine of ten geography courses in Arkansas were centered around trade and economics; in Missouri the most popular text on the eve of the war was still Whitbeck's *Economic Geography*. Soon, however, the political and cartographic dimensions of geography began to gain.[36] Nebraska high school geography courses tripled between 1942 and 1946, and the fastest-growing form of the subject—"world geography"—had become eight times as popular as economic geography. Most world and global geography courses were offered in the first two years of high school, while juniors and seniors were offered courses—by then considered common—in industrial, commercial, and economic geography. By 1946, 76 percent of the state's schools had fulfilled the State Department of Education's recommendation that a full year of geographical study be required of all high school students. In 1924, the only form of geography recommended in Massachusetts had been a commercial course for non-college-bound students, but all high schools expanded their geography course offerings as a result of the war and found students responsive. In Michigan, economic and commercial geographies continued to be popular, but by the end of the decade were competing with world geography courses. At the same time, California high schools were moving toward world and global courses in geography.[37]

The laments about geographic ignorance in the 1940s echoed those just after World War I. But the experience of the 1940s was in other respects unique, for what was now required was a reconceptualization of space itself. This renewed attention to the explicitly spatial dimension of world geography revived the teaching of geography in the schools and also made geography a remarkably popular pastime in the 1940s, as argued in chapter 9. The decade was, paradoxically, one of decline as well as advancement for geographical knowledge.

To understand the new direction of school geography requires a brief overview of the culture of geography generally. The demand for geographic expertise during the war had created a fissure within the profession during the 1940s. A growing pool of professionally trained geographers had long found themselves excluded from full membership in their discipline's professional organization, the Association of American Geographers. The AAG had allowed many of these young geographers to meet under its auspices, but prevented them from participating fully in its intellectual or institutional business. The AAG had prided itself on its limited membership and high admission standards, but this seemed particularly intolerable to a younger generation whose talents were so highly valued by the government during the war. Many younger geographers also felt that the organization had championed exclusivity in order to create a reputation for strict standards that would boost its image among neighboring disciplines. Those younger geographers who had served in the war found government work more attractive and lucrative than comparable academic posts, and thus there was not a dramatic return to college campuses from Washington, D.C.

after the war, as there had been after World War I. These circumstances led to the younger geographers' formation of the American Society for Professional Geographers, which they hoped would soon rival the AAG. F. Webster McBryde, the most vehement critic of the AAG in the 1940s, argued that the organization was mired in an elitist, stubborn mindset that reflected geographers' institutional vulnerability and self-consciousness. Targeting the profession's interwar history, McBryde argued that geographers'

> binding tenet was keeping their vaguely broad field of geography on a "mature," lofty, scholarly plane, far removed from the popular level of the *National Geographic Magazine*, as demanded by their physiographer-founder, W. M. Davis, while maintaining for themselves a prestige based on elitism. This was not professionalism in the true sense, for their strict, absolute requirement for membership was simply to have made an original contribution to "some branch of geography," a broad qualification indeed. . . . Anyone not complying to the satisfaction of the Council, personally and technically, faced the secret ballot and a freely-used blackball.[38]

McBryde's personal and professional frustration reveals the generational divisions plaguing the profession. The young members of the ASPG made their organization unique by limiting membership to those who had worked as professional geographers, most often in service to the federal government during the war.

In 1948 the contentious rift between the two organizations was healed when the ASPG merged with the AAG. But in that same year trouble erupted again when Harvard University abruptly announced the termination of its geography program. Though it had never been an independent department, geography's vitality under William Morris Davis in Cambridge was an important symbol of the discipline's early development, and a source of pride for the profession. The decision to close the program was a major defeat for geographers. Explanations for the decision are wide-ranging, but most agree that central to the crisis of geography at Harvard was the general intellectual vacuum that had developed after the interwar decline of environmentalist frameworks. When Harvard reviewed the status of the program in 1949 and 1950, no geographer could sufficiently defend the discipline to the university as a whole. When asked to define geography's intellectual domain, Edward Ullman—a geographer who sat on the university's review committee—defended the subject by outlining not what geography *was*, but rather what it was *not*: geology, environmental determinism, the study of place names, the preparation of guide books, or the work of the *National Geographic Magazine*.[39] This response seemed to exemplify the perceived conceptual weakness at the heart of the field. Harvard's subsequent judgment of geography as an unfit academic discipline was taken as a cue by many university administrators, and soon thereafter geography programs shut down at Stanford and Yale as well as a host of smaller institutions around the country.

But if the decade was generally one of turmoil for professional geographers, the public's frenzied interest in "geopolitics" represented something altogether different—namely, the subject's power to translate major social and political upheavals into coherent narratives. The emergence of geopolitics may have distorted the realities of international relations during and after the war, but its popularity reflected the need to understand the implications of a global war and far-reaching technological advancements. As it evolved, geopolitics meant many things to many people. Its original manifestation as German strategy—*Geopolitik*—was attributed to geographer Karl Haushofer, wrongly credited as the mastermind behind Hitler's concept and policy of *Lebensraum*. In a number of mass circulation magazines, beginning with the November 1939 issue of *Life*, Haushofer was characterized as the guru of geopolitics, and the key figure to be understood if the Nazis were to be defeated. As Gearóid Ó Tuathail has argued, though they contained serious misreadings of Nazi foreign policy, the lure of these pieces was their suggestion that the enemy was a rational and comprehensible one.[40] But there was also an element of mystery in these analyses, which detailed the concepts of political space supposedly employed by the Nazis but also granted the enemy a kind of omniscience, the power of "hidden logic" that was contained in the study of geopolitics.[41]

One observer has suggested that the study of geopolitics in the United States was not just an attempt to profile the enemy, but also a manifestation of America's own transition to international stewardship. Indeed, German *Geopolitik* was only an introduction to the fad. Its urgency motivated scholars, journalists, and freelance writers to search for similar sources of knowledge and power among American strategists. This accounted for the remarkable resurgence of Halford Mackinder's ideas in the 1940s, which many hoped would provide clues to political and strategic success. In 1904 Mackinder had declared the Russian landmass the "pivotal" source of international power in the future, a basis from which defensive and offensive stragies could be securely launched. To Mackinder, the world in the new century was increasingly governed by relationships over land. Mackinder's theories were made even more relevant by the onset of the cold war, when his idea of the Russian land mass as the "heartland" of world power took on a renewed legitimacy (fig. 6.1). Conversely, Alfred Thayer Mahan—whose work influenced Theodore Roosevelt's decision to strength the American navy—argued that international power would be won by control of the sea. In the 1940s a new generation of futurists announced that it was neither land nor sea, but air that would become the crucial contested space in the coming age of geopolitics. All three were admittedly reductionist, but each offered elegant and compelling perspectives on the complex realities of international relations.

Both German *Geopolitik* and the ideas of Mackinder were extensively circulated in *Newsweek, Reader's Digest, Life, Current History,* and *The New Republic* during the war. They were also vividly identified as a source of Nazi geopolitical omniscience in *Why We Fight,* the wartime newsreel series directed by Frank Capra for the War Department and the Signal Corps. In describing the German invasions early in the war, Capra used a map of the world centered on the North Pole to illustrate the proximity of the Americans to the European enemy, and then quoted Mackinder's geopolitical concept of the world island as central to Hitler's expansionist vision.

> Who rules East Europe commands the Heartland;
> Who rules the Heartland commands the World Island;
> Who rules the World Island commands the World.[42]

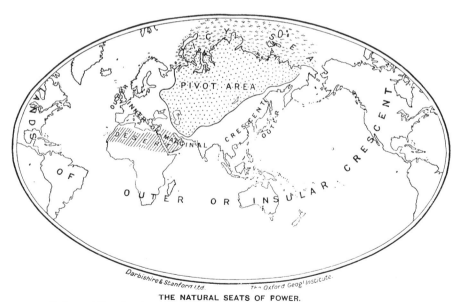

THE NATURAL SEATS OF POWER.
Pivot area—wholly continental.   Outer crescent—wholly oceanic.   Inner crescent—partly continental, partly oceanic.

Figure 6.1   Halford Mackinder's 1904 map illustrating "the natural seats of power." From Mackinder, "The Geographic Pivot of History," *Geographical Journal* (1904).

Mackinder's rule of land power made a powerful accompaniment to "Hitler's dream." His ideas also became integral to school texts and maps, as well as innumerable analyses of international relations in the earliest stages of the cold war.[43] By the end of World War II, geopolitics referred to a hardheaded, realistic cold war foreign policy, when in fact the ideas behind it never amounted to a coherent body of knowledge.[44]

One of the most influential advocates of this graphical reorientation for the public and for American school geography was George Renner. A teacher at Columbia Teacher's College and a consultant to both the Civilian Aeronautics Administration and Rand McNally, Renner spoke frankly of geography as a path to power. In one of his many wartime writings, he invoked the geographical sophistication of the Nazis as a way to demand a more political and strategic reading of the world in American schools. That Nazi Germany had a wholly different, more scientific and accurate understanding of geography suggested that Americans not only had to learn geography, but had to reimagine the world altogether. While most geographers and cartographers considered Renner hyperbolic and alarmist, more than a few educators took him seriously for his ability to construct a view of the world that reflected the nation's changing role abroad.[45]

Renner was influential for his use of geography to explain the world. Like Mackinder at the turn of the century, he spoke to the public by connecting the past to the future through the environment, and argued that historical and technological change had fundamentally altered the experience of space. Like the telegraph, the railroad, and the communication cables that had preceded it, Renner insisted, the advent of flight had shrunk distances. Invoking both Turner and Mackinder, Renner argued that at mid-century Americans faced a "closed system," which demanded an improved understanding of cartographic realities. Adhering to a conception of the world guided by the Mercator projection—which encouraged Americans to visualize themselves as largely separated from the rest of the world by vast oceans—was not just outdated by dangerous (see figs. 1.1 and 8.4, Mercator projection maps of the world). What Renner considered foreign relations failures—such as the rejection of the League of Nations—could be traced to a false sense of geographical isolation perpetuated by the Mercator projection. And to Renner's mind, a retreat into isolationism would be an attempt "to turn back the hands of the historico-geographical clock of world events and relationships."

Little wonder, then, that one of Renner's clearest legacies was the proliferation of new maps, both in textbooks and for general classroom use. Renner himself was partial to maps that graphically demonstrated what he saw as the "future" of global relationships. He brought Mackinder's maps of "heartland" and "periphery" back into discussion through textbooks, updating them so they were centered not on the Eurasian land mass but on the North Pole (figs. 6.2 and 6.3).[46] Schools were the ideal audience for this reorientation of geography. With "distances that are measured in hours instead of days," Rand McNally echoed Renner's warning that the enemy possessed a superior grasp of new geographic realities, or "geopolitics." Positioning itself as uniquely qualified to remedy this deficiency, Rand McNally reminded educators that enemy dictatorships had well-established courses in air education that the United States had only begun to replicate. Rand McNally's "World Map for the Air Age"—edited by none other than Renner—was centered on the North Pole in order to highlight the proximity of belligerents made real by aviation, and sold more in its first year of publication than any other map in the company's long history.[47]

Significantly, during the war Rand McNally sold one series of physical school maps and two series of physical-political maps, but marketed multiple political, historical, current interest, and air-education map series. The company also introduced its splashy new "air globe" during the war, a twelve-inch transparent ball that bore only the names of the important places in the world, omitting all information made "irrelevant" by aviation, such as lines of latitude and longitude, mountains, oceans, and even national borders. This air globe rested in a cradle rather than on an axis, so that Americans could study and view the world more truthfully, from all angles, and rid themselves of "antiquated" notions of hemispheric isolation generated by overexposure to the Mercator projection. In a carefully chosen phrase, Rand McNally and American Airlines marketed the globe to teach "the concept of freedom of the air and the universality of air transportation."[48]

The new polar maps were usually the first students would find in the text.[49] Most all were accompanied by long passages, even entire chapters, explaining the reorientation of cartography. In West, Meredith, and Wesley's *Contemporary Problems Here and Abroad* (1947), fifteen pages in the first part of the book were devoted to the new maps and images of the "air age" circulating through popular culture. *Air Age Geography and Society* (1945) had a section on "our global earth" that emphasized the new cartography as superior to Mercator's world. Grace Hankins devoted nearly a third of *Our Global World* (1944) to the need to adjust our concepts of time, space, and distance in the age of flight. To do this she borrowed the wildly popular "perspective" maps from *Life, Fortune,* and other serials, images that brought home to students the centrality of the Pacific and drew them toward

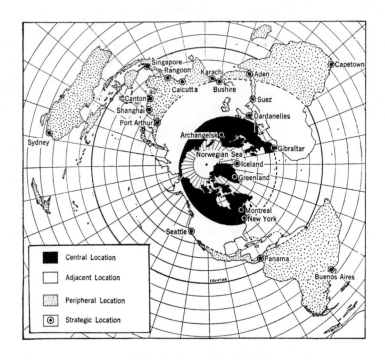

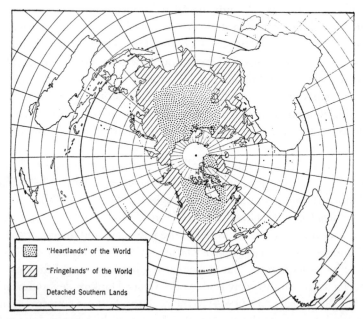

Figures 6.2 and 6.3 George Renner's adaptation of Mackinder's "heartland" thesis to the new "natural seat of power" at mid-century, the area around the North Pole. From Renner, *Human Geography for the Air Age* (1942).

a geopolitical perspective (see figs. 9.6–9.9). Predictably, these texts made the Mercator projection an enemy of truth and an egregious misrepresentation of space in the age of flight. By contrast, the new drawings and maps became paths to insight that gave the student a sense of discovery by viewing the earth from multiple angles (fig. 6.4).

Ironically, the greatest obstacles to Renner's vision of a population educated in the new strategies of the air age were his own colleagues. Geographers had spent much of the interwar period struggling to shed their discipline's intellectually anemic reputation, and thus avoided political implications of their work in order to concentrate on what might be considered "safer" fields of regional and economic geography. Interestingly, it was the geographers most successful in appealing to the public's interest—Huntington and Renner, for example—who caused the profession the most embarrassment. Though academic geographers shunned Renner's geopolitical speculations and environmentalist leanings, many other textbook authors spoke to this renewed interest in political geography and geopolitics. Packard, Overton, and Wood concluded *Geography of the World* by exploring modern politics, the spread of nationalism, and the potential problems this posed to political, social, and geographical diversity. Only certain nations were included: Japan, China, Russia, Italy, Germany, France, Sweden, Britain and her empire, Chile, Argentina, Brazil, Mexico, the islands of the Pacific, and the United States. Within this "circle of interest" in the geography textbooks, one can speculate even further about a new sense of awareness of certain parts of the world. It is important not to make too much of these trends: an emphasis in textbooks does not correspond to public awareness. Yet it is remarkable to see in the texts a consistent emphasis on certain areas of the world. The case of the Pacific is instructive.

In 1898, the Philippines, Guam, and Wake Island gave the United States an interest in the Pacific as a "gateway to the Orient." But throughout the first half of the twentieth century textbooks treated the Pacific, or Oceania, as a distant group of scattered islands, a buffer *between* civilizations more than a society of its own. By contrast, the importance of the Pacific in World War II turned the public's attention toward it as an integrated, relevant region. The attack on Pearl Harbor jolted the nation out of its sense of relative invulnerability to the west. The Mercator projection, usually placing the Americas on the western half of the map, made the Atlantic the central body of water between the hemispheres. The war, however, forced the Pacific into the public's mind, and most of the texts of the decade paid careful attention to the region, not just as a place of leisure but as a critical space of its own. Textbook authors scorned the ignorance of

Figure 6.4 Map display in the foyer of Brooklyn's Bay Ridge High School in 1942. Notice especially the multiplicity of cartographic perspectives, including the polar projection umbrella in the left foreground. New York City Board of Education Archives, Milbank Memorial Library, Teachers College, Columbia University.

the Pacific that had lulled the nation into a false and dangerous sense of hemispheric security. The American possessions in the Pacific were cast as steppingstones to Asia, first aiding the U.S. war effort against the Japanese and then allowing a path to Eastern markets.[50] Chamberlain called them our "new neighbors," brought closer by the use of the airplane and the lessons of the war. Correspondingly, as argued in chapter 9, the atlases of the 1940s began to draw maps illustrating the importance of the Pacific as a body of water joining, rather than separating, the continents on either side (see fig. 9.15).

GEOGRAPHY AND THE FAMILY OF MAN

This imperative of treating what had been distant lands as newfound neighbors created a striking strain of universalism in wartime textbooks. On close inspection this tendency indicates a pronounced departure from earlier forms of environmental determinism. The real enemy of the mid-century textbooks was not the savagery of the Africans or the indolence of the Spanish, but the persistence of ignorance and fear that was visually embodied by the Mercator projection. Narratively too the textbooks identified threats to human progress that were conquerable, and in the process created a world that was potentially conflict-free: cultural, ideological, or economic misperceptions and differences could be solved through an understanding of humanity's shared needs. While the late nineteenth-century texts described a static, rigid place of irreconcilable differences, the mid-twentieth-century texts found a fluid landscape where people shared not just an interest in economic advancement, but basic human instincts for love and cooperation.[51]

This was part of a larger trend at mid-century, what David Hollinger has described as "the well meaning but unsuccessful efforts to address the species as a whole." The horrors of Nazism and the threat of the cold war encouraged some to assert that humanity were driven by shared Enlightenment notions of knowledge and rights. Consider Edward Steichen's highly successful photographic exhibit—"The Family of Man"—designed to highlight the fundamental dimensions of human experience, irrespective of history, geography, or culture.[52] This universalism was encouraged by the social studies, which cast education as a form of problem-solving. Mid-century texts were preoccupied with concepts and the problem method, dismissing the recitation of facts that had so thoroughly characterized the educational models of the nineteenth century. One geography text asked "What can we learn from the Egyptians?" and "Why did Rome fall?" Contrast this with typical questions from the nineteenth- and early twentieth-century texts, such as "Bound Austria," or "What are the chief classes of the animal kingdom?" If nineteenth-century school geography catalogued the world, twentieth-century geography explained how it worked.

After World War I, the moral force of the physical environment and of race slowly gave way to different descriptions and explanations of relative progress. Rooted in a kind of universalism, these modern texts were eventually purged of the moral and judgmental force of climate and race as an influence over human behavior. In Chamberlain's *Air-Age Geography and Society*, a culture that would once have been described as indolent or savage was now conceived of differently.

> When a few hours' work provides for his needs, the Polynesian sees no reason in working longer. It is not that he is lazy. He is capable of prodigious work—when he sees some point in doing so. But he works to live, not lives to work. The idea of working to accumulate wealth is foreign to him. Consequently, steady routine jobs have no attraction and plantation labor is usually brought in from other areas, especially from India and China.[53]

Throughout this passage are traces of contemporary anthropology, approaching human difference as a product of relative values and diverse cultural contexts, all of which are minor next to the essential unity of the species. These texts celebrated difference not as a rigid product of environment and race but simply as similar modes of communication situated in different cultural contexts. Meyer and Hamer captured this sentiment in a chapter that explained the diversity of customs around the world as simply different expressions of the universal "habits of man." Mid-century texts sought to celebrate the universal over the particular, all the more in an age threatened more by ideological than by racial differences.[54]

This new approach matched the reoriented narrative of contemporary atlases. In a special section added to Hammond's 1947 *New World Atlas*, and included as well in the 1950 *Library World Atlas*, readers found a review of the "Races of Mankind." The essay identified three main races—Negro, White, Mongoloid—then examined the range of diversity within these groups in order to address the current interest in international cultures sparked by the return of globetrotting servicemen. The author, an anthropologist, used the category of race to emphasize not the hierarchy of difference that was so central to late nineteenth-century atlases and textbooks, but rather the unity of mankind.

> One of the inevitable consequences of the global war is a wider knowledge concerning the peoples that inhabit the world. Places

and human tribes that have before been unknown to the masses, come into closer perspective. The men of the armed forces bring back vivid and often grateful memories of distant lands and natives who have shown how brave and trustworthy primitive tribesmen can be. When we live with these children of the jungle, or the desert, we are amazed to find them so full of humor and natural intelligence, so human. Their code of honor is often a challenge to the more highly educated peoples. Their word is better than a written agreement—it cannot be torn up or ignored. . . . Until we learn to appreciate the good qualities of our primitive fellow men, and stop preaching about merely tolerating them, we will find it difficult to answer satisfactorily this searching challenge.[55]

The piece went on to criticize the misuse of the word "race" by contemporary anthropologists:

["Race"] has even been made to include psychological and social qualities, and has often been confused with nationality; also, it has been allied with ideas of inherent superiority or inferiority. Let it be clear that by "race" we mean a certain combination of heritable physical traits, without any implication of social status or psychological attributes.[56]

Race had come full circle: in the late nineteenth-century atlases and school texts it was used to separate and classify, but by the 1940s it had become a self-conscious category used to underscore the "brotherhood" among peoples in a world now divided by economic and political ideology. In the process, the atlases and textbooks had shed their nineteenth-century perspective.

In the first two decades of the twentieth century, Rand McNally's atlases opened with a two-page sketch of the "Occident" and the "Orient," physically separating the people, geography, architecture, and lifestyles of the East from those of the West. But by the 1940s the direct threat of war made geography unavoidably spatial, thereby bringing the United States closer to the rest of the world and replacing its hemispheric, regional identity with an international one. The postwar atlases and textbooks reflected this change by presenting the world as qualitatively more *interdependent*. By organizing the world according to geopolitical regions rather than strict state boundaries, they emphasized the geographical situation. The introduction of topographical maps of the continents further reinforced this perspective by replacing borders with mountain ranges and lowlands, which depicted a seamless landscape and permeable national boundaries. In the cold war, race became incidental to political and economic association, and a source of unity rather than division in atlases and textbooks. Correspondingly, though texts and atlases still referred to race and climate, they emphasized the degree to which races adjusted to and mastered their natural environments. The advent of aviation encouraged this view of the environment as plastic and subject to human ingenuity. Anticipating a world of air travel made formidable elements of the physical environment—deserts, oceans, and glaciers—seem less relevant. Even continents, previously understood as the fundamental building blocks of geography, disappeared from some globes that marked only cities.

Yet despite these changes there were at mid-century still clear traces of earlier eras. Africa was still, to Meyer and Hamer, the "dark continent," described in terms of South Africa's mineral wealth, Livingstone's benevolence, and Albert Schweitzer's medical mission. Students were asked to explain why Africa was known as the "continent of colonies," and which of these were most valuable. These authors celebrated racial difference at the same time that they retained older ideas of civilization, barbarism, and savagery, now defined largely according to the ability "to *master* the land."[57] It was the language that had changed: one text purported to show "the extent to which man has brought himself and his environment into better adjustment and, in so doing, has improved his economic welfare." Thus the following passage describing the Chinese, inconceivable in 1880, 1900, or 1920, was by 1948 quite ordinary:

> The favorite delicacies of the Chinese are not rats and snakes.
> They eat other things than rice. All the people are not laundry-
> men. All Chinese do not look alike. All Chinese are not cunning
> and crafty. They have souls even if they are not all Christians.
> They are not a mysterious and inscrutable race and they do not do
> everything backwards. We should not think of them as queer, but
> only different.[58]

Driving home the message, this passage was accompanied by a photograph of the Chinese, not engaging in exotic rituals, but playing games similar to those of Americans. Another text approached China in a similar spirit, hopeful that common ground could be found with a nation that had so admirably fought the Japanese. Photos of the Chinese on succeeding pages sympathetically portrayed their resistance to Japanese aggression as "a valiant attempt to preserve the traditions of their centuries-old civilization." Chamberlain identified the Chinese as potential allies who shared our values, and represented them as a poor but dignified, intelligent, and determined people.[59] On the eve of the revolution, relations with Nationalist China were still hopeful, and urgent, for this was a world divided not by racial exoticism or climatic belts but by economic and political ideology. The fluid place of China in these textbooks illustrates the degree to which school geography was malleable, influenced not just by increased scientific knowledge but by the rise of American stewardship abroad and the transformation of educational imperatives at home.

7

Negotiating Success at the *National Geographic*

1900–1929

In early September 1909, two different Americans declared themselves the first to reach the North Pole. On September 1, a little-known Brooklyn physician named Frederick Cook announced that he had arrived at the top of the world in spring 1908. A few days later, the renowned explorer Robert Peary dismissed Cook's announcement and instead claimed that *his* was the only true discovery of the Pole, made in April 1909. By this time, the National Geographic Society had cultivated a relationship with Peary for over a decade, funding his expeditions and honoring him with the Hubbard Medal in 1906 for "arctic explorations farthest north." His prior expeditions had earned him esteem by the Society and the general scientific community, while Cook had no such reputation, and in fact was hounded by charges of fraudulent financial dealings. When the two competing claims caused a scandal that gripped public attention and divided the nation, the National Geographic Society assumed the mantle of scientific authority by asking both men to submit their records to a committee of the organization's officers for verification. After a relatively superficial investigation the committee verified Peary's records, and effectively nullified Cook's claim by simply ignoring it. William Morris Davis—longtime associate of the Society and an eminent geographer at Harvard—remarked to committee member and Society president Henry Gannett that "Cook needs examination by a criminological psychologist rather than by geographers."[1] That December, at its annual dinner, the Society honored Peary, not Cook, as the true conqueror of the North Pole. Despite the lack of conclusive evidence—neither's records can be entirely authenticated—the Society vociferously and publicly endorsed Peary's expedition over Cook's.[2] In light of the Society's long and close relationship with

NEGOTIATING SUCCESS AT THE *NATIONAL GEOGRAPHIC*   149

Peary, its partisan actions raised more than a few eyebrows, though pivotal figures such as former President Roosevelt lent their immediate support to the decision.

The Cook-Peary feud was tailored to the needs and purposes of the National Geographic Society in the early twentieth century. Both a gripping story and a thorny scientific problem, the controversy allowed the Society to assume several roles, all of which offered it the opportunity for self-aggrandizement. As an organization that had long supported Peary's efforts, the Society positioned itself as advancing both the noble goals of science and the heady dreams of exploration. As the self-proclaimed arbiter of the dispute — preempting a Congressional investigation — the Society asserted its position as a sober scientific authority. But, as Davis's comment indicates, the Society was also engaged in just the kind of mudslinging it publicly disdained, and as an organization was as fallible as any other. The *National Geographic Magazine*'s ongoing coverage of the North Pole saga gained it prestige as well as notoriety, and fueled the increase in its membership prior to World War I.

The phenomenal growth of the National Geographic Society in the 1910s and 1920s is all the more compelling given its internal contradictions. It held nonprofit status until 1997, yet was an organization worth millions by the 1920s. It identified itself as an elite scientific society, yet refrained from "technical" discussions that might alienate those who were generally "ignorant" of geography. The Society's journal — the *Geographic* — strove to be timely and to address current issues, yet also to be of permanent value, and boasted of the public's treatment of the monthly as a book rather than a disposable magazine. The Society identified its mission as both national and cosmopolitan. It promised readers a scrupulously accurate profile of "the world and all that is in it" that would — paradoxically — abstain from anything of a "critical or controversial nature." It made a policy of avoiding political controversy and showing only the bright side of life, a policy that eventually led to the celebration of progress and change in Fascist Italy and Nazi Germany in the 1930s. It stood as the guardian of propriety and wholesome entertainment, but regularly and unapologetically printed photographs of nude women from exotic lands, a practice that shocked some and lured others. In part, the magazine succeeded beyond all expectations because it balanced, or rather *appeared* to balance, these contradictory goals, and in doing so appeared to be all things to everyone. Despite, or perhaps because of, these apparent contradictions, the National Geographic Society managed to build a membership that reached one million by 1926 and continued to increase through the rest of the century. A close look at the *Geographic*'s maturation in the early twentieth century reveals how the magazine finessed these contradictions and negotiated the expectations of its members, in the process communicating ideas about the world to an ever larger and more diverse audience. This chapter examines the evolution of the magazine in the early twentieth century; the Society's maps are treated in chapters 8 and 9.

As suggested in chapter 3, the modern form of the *National Geographic* is grounded in the expansionist turn taken by the United States in 1898. The Spanish-American War helped the magazine evolve from a technical journal of natural science to an illustrated magazine of current, political, and human interest. This change also fed the magazine's incipient use of photography. After 1900, the editors began to institutionalize what had previously been an experimental approach, and set policies they hoped would build the magazine's readership and reputation. In 1905 the Society gained its largest proportional increase in membership, and later bursts of growth also came before and after—though notably not during—the nation's involvement in World War I. Members proudly displayed the magazine in their homes as a marker of sophistication, and showered the Society with earnest testimonials such as "I would rather be without shoes than this magazine."[3] This kind of loyalty suggests that the *Geographic* was not just a popular magazine, but an institution that resonated deeply with American culture. In other words, to understand the Society requires a consideration of what it meant to profess, as one member did in 1923, that "I do believe in the *National Geographic*. I certainly do."[4]

THE NATIONAL GEOGRAPHIC SOCIETY IN AMERICAN SCHOLARSHIP

Recently the *Geographic* has drawn the attention of scholars interested in its racial and ideological orientation. Catherine Lutz and Jane Collins argue that the magazine grew out of a kind of progressive evolutionism that assumed the superiority of Western ideas, but also expressed a buoyant attitude about less industrialized societies. This posture distinguished the *Geographic* from the cruder forms of Social Darwinism typical of the contemporary nativist and eugenics movement in the first two decades of the century. The Society, they argue, adopted a different strain of evolutionary thought, one that "focused on the 'evolutionary guarantee' of progress through the increasing triumph of rationality over instinct even as it continued to justify residual inequalities of sex, class, and race. *National Geographic* reinforced America's vision of its newly ascendant place in the world by showing 'how far we've come.'"[5]

The assertion that neither nativism nor eugenics had currency at the

Society is difficult to accept. In fact, prior to World War I the *Geographic's* contributors and editors adopted nativist and eugenicist arguments when advocating restrictions on immigration. Yet Lutz and Collins are right to suggest that the *Geographic's* sensibility was complex. The magazine's development in the twentieth century into a heavily illustrated and popular magazine made the editors much more self-conscious about their work, which explains their—admittedly unsuccessful—attempts to transcend the ideological currents of the day. Given this observation, the critical question becomes how these ideals actually played out in the pages of the magazine. Whether the "progressive" evolutionism Lutz and Collins identify worked to serve or to complicate the *Geographic's* mission, and whether this approach was resisted or contested among the editorial leadership, is worth exploring.[6]

Geographer Tamar Rothenberg draws from cultural theory to grapple with some of these questions. Borrowing concepts from Mary Louise Pratt, Rothenberg attributes the *Geographic's* success to a "strategy of innocence" where editors and contributors championed their work as politically neutral while at the same time embracing the imperatives of European global authority. While Lutz and Collins observed a unique variant of optimism pervading the *Geographic*, Rothenberg identifies the persistence of an "anticonquest narrative," which she defines as "a simultaneous assertion of innocence regarding the larger imperial project and complicity with that project."[7] I would go even further, and argue that the *Geographic's* enduring appeal had much to do with its posture of political neutrality and scientific authority, where "interest politics" were dismissed even as the Society touted the virtues of American interventionism. Yet while Rothenberg's "strategy of innocence" nicely frames the dominant ideal for the *Geographic* in the early part of the century, the magazine almost invariably fell short of this elusive goal, and if we are to understand the contradictions within the *Geographic*, a more explicit comparison of the ideal and the actual magazine is needed. Though these recent accounts have brought a good measure of critical theory to our understanding of the *Geographic*, frequently they insufficiently develop its historical context. For example, most accounts fail to assess properly the Society's experience during World War I, a time of abrupt and significant stagnation rather than growth, as is commonly assumed.[8]

Furthermore, any study of the early *Geographic* must take account of its relationship not just to the war, but also to contemporary magazine culture. During this decade the literary and current events magazines that had dominated since the Civil War—such as *Scribner's Monthly*, the *Century*, the *Atlantic Monthly*, and *Harper's*—were gradually eclipsed by a new generation of less expensive magazines geared to different tastes and concerns. Three of these—*Munsey's, Cosmopolitan,* and *McClure's*—were far more dependent on advertising than on subscriptions, were markedly more visual than their predecessors, and reached a far wider audience. Recent scholars have paid much attention to the meaning of this transition. Richard Ohmann argues that the cheap monthlies of the 1890s played a critical role in the rise of mass culture and the development of consumer capitalism. Filled with highly visual and persuasive advertising, these magazines helped Americans adjust to a new array of goods and services. Magazines such as *Munsey's* sold at a loss in order to boost circulation and attract advertisers, and thus to Ohmann symbolize a critical shift from a production to a consumption economy.[9]

Though Ohmann's study is controversial for its Marxist approach to culture, other scholars have agreed that the magazine revolution of the 1890s marks a significant historical moment. While concluding with Ohmann that these new popular magazines were central to the development of corporate capitalist culture, Matthew Schneirov insists that we take seriously the progressive vision of their editors and authors.[10] To Schneirov, the reformist zeal of these progressives turned their magazines into a force for social change, a function as consequential as their role as agents of capitalist hegemony. By shaping middle-class values and sentiments, he insists, these magazines helped redefine culture itself from a "repository of timeless and permanent virtues" that looked to the past for inspiration into "a series of future projects or 'dreams'" that arched forward.[11] Schneirov assumes that magazines had the power to effect social and cultural change. By contrast, Ohmann considers modern mass-market magazines important largely as vehicles for advertising, a function that eventually came to control the medium itself. While these two scholars use different approaches—Schneirov a cultural lens and Ohmann a materialist one—both point to the turn of the century as a pivotal moment in the history of American magazines specifically and print culture generally.

The *Geographic* occupied an odd place in this transition. It gazed into the future and drew strength from the past. In the 1890s the *Geographic,* with its limited circulation, technical content, and paltry advertising, could not be defined as popular. Only toward the turn of the century did the editors begin to think of the magazine as having a potentially national audience. In terms of style it was closer to *Harper's* textual focus than to *Munsey's* visual appeal. In other words, the structure and goals of the *Geographic* distanced it from the magazine revolution. Yet, as argued in chapter 3, the Spanish-American War created a climate of experimentation that turned the magazine toward the human, the political, and the visual. By the end of the First World War, photographs dominated the *Geographic*, and it had accumulated a solid list of advertisers, though never as extensive as those of the "new" magazines such as *Munsey's*, which averaged 73 pages of ads per issue in 1895. Revenues generated from this level of advertising allowed the publisher to price *Munsey's* at only ten cents an issue, less than one third the cost of *Harper's* and less than half the cost of the *Geographic*. From its origin through 1919 the price of each *Geographic* remained twenty-five cents, even though the amount of advertising grew dramatically after 1910. The *Geographic* carried 39 pages of ads in May 1915, 63 in 1920, and 80 in 1926. In the early decades the magazine's limited advertising was dominated by railroad promotions, perhaps on the assumption that those with an interest in foreign lands might also be inclined to travel domestically. In the 1910s the *Geographic* carried ads for books, typewriters, cameras and other luxury items, automobiles, investment houses, and insurance companies.[12] Despite its relatively high price and late arrival as a mass-market magazine, the *Geographic* came to exemplify this new, intensely visual genre of middlebrow publications. By 1919 its impressive circulation was second only to that of the *Saturday Evening Post*, and exceeded the combined distribution of the *Atlantic Monthly*, the *Century*, *Harper's Magazine*, *Outlook*, *Review of Reviews*, *Scribner's*, and *World's Work*.[13] Thus the *Geographic* outstripped the circulation of the venerable older monthlies and the newer mass-market magazines even though it identified as a journal for members of an elite scientific Society. A closer examination of just what it meant to be a member in this elite Society goes a long way toward explaining the *Geographic's* improbable success.

"Mention the Geographic—It Identifies You"

On each page of advertising in the magazine readers were reminded to "mention the *Geographic*—it identifies you." Though on one level simply an attempt to promote advertising with its readers, the quip carried an added meaning at the *Geographic* because for most of the century the magazine was available only to "members" of the Society, a distinction that was more than semantic. In fact, only in recent years has the Society sold the *Geographic* on newsstands to the general public. The idea of membership was rooted in the nineteenth century, when the Society was a scientific body and the *Geographic* the organ of its proceedings. After 1900, however, the designation of "member" in an organization that grew by tens of thousands each year seemed misleading given that someone who could not find a sponsor would be provided one. In fact, some older members—those with more academic, technical, and scientific training—attempted in 1899 to separate the magazine from the Society, to sell the former on the newsstand to the general public, and to maintain a selective membership for the Society itself in Washington. Though the plan was abandoned, it suggests that "membership" was understood as a nominal category by some as early as 1900. Why, then, did Alexander Graham Bell—then the Society's president—insist that the system of membership be maintained even in the face of strong internal opposition?

Bell believed that the obstacle of membership actually enhanced the desirability of the Society, and he hoped to deepen the appeal by publicizing its selectivity. Here again the success of the *Geographic* grows from a contradiction. By maintaining the category of membership the Society invoked a sense of elitism, and continued the tradition, if only in name, of the amateur geographical and scientific societies that had flourished in the nineteenth century. Yet at the same time Bell made perfectly clear that the content of the magazine would be accessible to all its readers, who by his own admission "joined the Society on account of their ignorance of geography, not their knowledge." The *Geographic* would serve the mass-market tastes of the twentieth century through a yellow-bordered monthly that radiated nineteenth-century cultivation and privilege.[14]

Were it not so successful, the category of membership would be unremarkable, judged as a self-serving pretense or a crass marketing technique. But by all accounts, members took the distinction seriously. As Margaret Wilson, a missionary with the Young Women's Christian Association, wrote in 1926, "I don't know when I have ever been so proud of anything as the little certificate which I hold as a member of the Society. It is second to my college diploma."[15] Her degree of enthusiasm was exceptional, but the actions of others echo her sentiment. As early as the 1910s, Americans accorded the *Geographic* special status in the home, collecting issues in the library—then the basement, attic, or garage—rather than discarding them along with other magazines. But what did the *Geographic* and Society membership actually mean? Some thought the magazines had "the best description of people and places of any that I have read" and considered them to be the "best library money can buy."[16] Others were struck by the *Geographic*'s adventurous appeal: one Midwesterner was thrilled to discover that "I can completely lose myself in its pages in most any country or climate I desire."[17] In 1923 an eighteen-year-old Minnesotan recalled that "when I first saw the *Geographic* some six or seven years ago, I stayed up until midnight perusing its magical pages."[18] Common to all was the sense that they were engaged in a shared and *respectable* enterprise, be it educational, scientific, adventurous, or entertaining. As Philip Pauly argues, the Society was in the unique position of being an amateur institution that survived—even flourished—in an increasingly professionalized culture.[19] This sense of membership in and ownership of the Society's work would have important implications for the public's understanding of what the *Geographic* ought to become.

While Bell deserves much credit for maintaining the category of membership, his most influential successor—Gilbert Hovey Grosvenor—used the idea not just to enhance the loyalty of readers but also to curry favor with advertisers. Because the Society cultivated a membership, it could claim knowledge of its audience, and thereby tailor its appeal to particular advertisers. Here the *Geographic* distinguished itself *through* its readers. Not all were wealthy, the Society conceded, but they were economically and socially stable, rooted in the American system, morally upright, cosmopolitan, and white. The *Geographic*, advertisers learned, never catered to "current frenzy, fad, or unrest,"[20] for its members, the "Upper Masses," were uninterested in such an approach:

> The Upper Masses are the million or more families whom Good Fortune, Intelligence, and Energy have put at the top . . . [with a] high standard of living which flourishes in wet and dry seasons because it is rooted in tradition . . . not all are wealthy . . . but they have incomes far above the average . . . assured by their native resourcefulness, versatility, varied assets, important friendships, community standing, intelligence and constructive energy.[21]

Such a statement bore indelibly the imprint of Grosvenor, editor of the *Geographic* by 1903 and president of the Society by 1920. As others have observed, Grosvenor was fastidious about the Society's reputation of respectability. Ironically, this editor, who was legendary even then for his photographs of barebreasted women, also canceled the advertising account of Listerine when the company refused to remove references to "bad breath" from its advertising copy in 1921. Grosvenor reported to the Society's Board of Managers with satisfaction that while the ad failed to meet the high standards of the *Geographic*, "the *Saturday Evening Post, Literary Digest*, and *Scribner's* passed similar copy without question."[22] Grosvenor's proud and satisfied tone came from his conviction that he spoke the same language—particularly in terms of propriety—as the Society's members, and thus spoke *for* them. But this idea of membership also meant that the magazine was more accountable to its readers. And as the membership skyrocketed—reaching 500,000 by 1916 and one million by 1926—the editors had to negotiate the content of the *Geographic* with increasing care.

CURIOSITY AND NATIONALISM IN THE
IDEAL OF THE *GEOGRAPHIC*

How did the *Geographic* make sense of the world in a way that so many found uplifting, educational, and entertaining? Having shepherded the Society through the excitement of the Spanish-American War, Bell was brimming with ideas about the magazine's future. He knew firsthand the challenges of magazine publishing, having edited *Science* at the time of its initial, failed incarnation in the 1880s. This led him to pay close attention to the appeal of the *Geographic*. Bell held up two contemporary monthlies as models for the successful general interest magazine: the *Century* and *McClure's*. As one of the most widely read monthlies of the 1870s and 1880s, the *Century* had a reputation for serious reporting and a taste for the reformist politics of civil service reform, conservation, anti-bossism, and the secret ballot. The retrospective surveys of the Civil War it published in the 1890s electrified its readership, as did George Kennan Sr.'s controversial chronicles of his Siberian journeys. The *Century's* mix of depth and breadth had managed to reach a wide readership, prompting Bell to write the following to Assistant Editor Grosvenor in 1899:

> Having in my mind an ideal, a magazine like the *Century*, of popular interest and yet scientifically reliable as a source of Geographic information we have something to work towards. Our efforts, therefore, should be to enlarge the magazine, have it bright and interesting, with a multitude of good illustrations and maps. . . . dry and long-winded articles of technical character should be avoided as much as possible.[23]

But while the *Century* had commanded wide respect in the nineteenth century, its circulation was hit hard by the magazine revolution of the 1890s, and thereafter began to decline. Out of this turbulent decade came the new ten-cent monthlies, the most serious and commanding of which was *McClure's*.[24]

With its low price, bright illustrations, and timely content, *McClure's* reached a circulation of 370,000 by 1900 (second only to *Munsey's*), while that of the *Century* fell that same year to about 125,000. The *Geographic's* membership trailed far behind both, hovering just above 2,000. Bell admired both the *Century* and *McClure's* for their ability to translate scientific news, such as expeditions, explorations, innovations, and discoveries, into language accessible to the lay reader. Americans were eager to understand the breakthroughs of Marconi and Pasteur, to follow the polar expeditions northward, and to see animals photographed in the African wild. Both magazines also chronicled the nation's work abroad in the 1890s. The political urgency of the Spanish-American War taught Bell to collect material on impending and current foreign problems, such as the Transvaal, the plague in Egypt, and the Russo-Japanese War.[25] When *Review of Reviews* republished articles from the *Geographic* in 1900, Bell was sure that the magazine had the potential to move beyond the capital to the nation at large. The focus of the *Geographic*, quite simply, would be in the "geography of current events."[26]

Alongside this focus on science and international relations, Bell reminded Grosvenor to appeal to the reader's imagination. Convinced that the public would have to be lured into reading geographical materials, Bell himself combed the issues of *Science, Nature,* and *Smithsonian* for curious, odd, and exotic items that might grab the attention of the average reader. Consequently he suggested that readers learn about the Russo-Japanese War through a profile of dysentery among Russian soldiers, and thought the geologic structure of the Aleutian Islands would be far more interesting if readers also discovered that, due to chronic earthquakes, the islands had actually sunk twenty to twenty-five feet. Bell also thought small oddities and discoveries would strike a chord with readers: that mosquito netting prevented malaria, that the Japanese waltzing mouse could be explained by a deficiency in the ear canal, that a kind of cannibalism had been observed in the ovaries of "humpbacked or deformed" (and therefore visually captivating) fish off the coast of South Africa.[27] Bell's *Geographic* would deliver world geography by appealing to human curiosity. As he put it:

> We don't want to abandon the principle of giving the public reliable and substantial information; we don't want to lower the scientific standard in any manner or degree of the facts we communicate. We don't want to cut off anything, but WE DO WANT to add on something—and that is in one word LIFE. I don't see why we shouldn't have anecdotes of a suitable kind, even jokes, although I am not quite so sure of the latter. . . . The WORLD is our theme; but mountains and valleys and the sea and the air are— by themselves—uninteresting things. There must be a *reductio ad hominem,*— or at least some reference to life in some of its forms. . . . We are not limited to geology, hydrography, or meteorology, but include within our sphere, *the world itself and all it holds.*[28]

Bell's observation that "mountains and valleys and the sea and the air are — by themselves — uninteresting things" spoke to the dilemma faced by contemporary educators, who were convinced that physical geography drew only limited student interest. After the turn of the century, teachers, school reformers, and professional geographers all began to gravitate toward geography as a study of the landscape in relation to *human* behavior. Grosvenor himself admitted this to be at the heart of the magazine's growth in 1905, commenting that "the renewed interest in geographic matters [is] due, apparently, to the recent additions of our territory, the expansion of our foreign commerce and commercial interests."[29] As argued in chapters 4 and 5, this in part explains the increasing focus on commerce within school geography after 1900, a framework that united the human and natural worlds and expanded the potential for geography. In the same way, Bell and then Grosvenor developed the *Geographic* to include, as much as possible, the human dimension of geography.

This vision for the magazine — appealing to the reader's sense of curiosity — rested upon the extensive use of photography. Even a cursory glance at the early twentieth-century *Geographic* bears this out: rarely are people absent from the illustrations. As argued in chapter 3, the *Geographic*'s experience of the Philippines and Cuba during the war with Spain had convinced Bell that the textual narrative would always be subordinate to the illustrations. Given his emphasis on human subjects, Bell enthusiastically endorsed the photographs of seminude African women in the March 1900 issue as precisely the type of illustration that would "bait" members into reading about South African culture in the text (see fig. 3.2).[30] Within a few months, barebreasted Bolivian women appeared in the *Geographic*, reflecting the editorial board's tacit acceptance of Bell's strategy. This was a turning point for the conception of the magazine, in terms of both form and content.[31]

With its increasing attention to human geography, the Society was deeply interested in explaining human behavior and history by reference to both race and environment during the first decade of the twentieth century. And, like academic and school geographers, the *Geographic* did not easily reconcile the apparent contradictions between the two, though it generally leaned toward references to the determinative power of race. As chief of the Weather Bureau — and future president of the Society — Willis Moore explained in 1904, "climate is the most potent of any factor in the environment of races. It is climate and soil plus heredity and form of government that produce either vigorous or weak peoples."[32] But a few years later, the degenerate state of Haiti was explained by the failure of the race to live up to the island's ideal geographic position and environment.

Miscegenation between Africans and southern Europeans created a population with "characters as low as can exist in human nature," prone to greed, avarice, cannibalism, and voodoo.

The discovery of Mendel's laws had fueled a general interest in eugenicist ideas and explanations, and these were especially appealing to those at the *Geographic* given the Society's keen interest in the problem of immigration. In an address to the Society in late 1904, the former Assistant Commissioner of Immigration comprehensively profiled the new immigration, calling for renewed attention to "racial geography" education in the schools and legislative restrictions to ensure the quality of those who arrived at Ellis Island. Grosvenor followed up this plea with a statistical demonstration that the ethnic groups most literate and least likely to commit crimes were in fact northern Europeans and Scandinavians.[33] A few years later, Grosvenor again endorsed proposed immigration legislation in the hopes that it would "debar imbeciles, weakminded and other undesirable classes." Bell's address to the American Breeders Association, reprinted in the *Geographic*, advocated the study of evolution and ethnic "types" in order to encourage those groups—though unspecified—that would improve the national stock and discourage those deemed harmful. The *Geographic*'s concern with immigration restriction perhaps peaked with Robert DeCourcey Ward's nativist preachings of 1912, arguing for the application of eugenicist principles to our understanding of immigration law. This early twentieth-century trend in the *Geographic* reflected less a "strategy of innocence" or progressive reform than one of overt racial hierarchy.

Consider the contributions of Theodore Roosevelt, perhaps the National Geographic Society's most honored member. In late 1910, after spending nearly a year in Africa, Roosevelt recounted the exciting details of his journey at the Society's annual dinner, his stories replete with extraordinary hunting feats and encounters with exotic native tribes. The *Geographic* subsequently published Roosevelt's adventures as "Wild Man and Wild Beast in Africa," abundantly illustrated with photographs taken by his son, Kermit (figs. 7.1 and 7.2). With pride, Roosevelt noted that not a single white man had been lost, though two natives had been mauled by animals, and a few killed by disease due to their inability to care for themselves. Their native guides and attendants, he recalled, "were like great big children . . . with no ability to think of the future." Roosevelt moved easily between descriptions of the natural and the human worlds, reinforcing the similarities between "wild man and wild beast." His skills as a raconteur were buttressed by his anthropological observations of cultural practices and racial identities of the Africans themselves.[34]

Kermit Roosevelt's photography—like the rest of the *Geographic*—

Figures 7.1 and 7.2   Photographs of Theodore Roosevelt's 1910 journey in east Africa, reprinted in the *Geographic*.

powerfully connected the human and natural worlds. As Tamar Rothenberg points out, the *Geographic* portrayed people as "types" in their natural environment, a practice that naturalized the group represented, and attributed characteristics to them as a whole, thereby reducing the collective to a singular, one-dimensional figure whose repeated appearance in the *Geographic* gradually becomes in itself the expected figure. To Rothenberg, this practice of "typification" helped entrench the American anti-conquest narrative by placing the subjects in "a visual narrative of place-bound cultural timelessness." This categorization lent further authority to the seer—the American—as the master of a physical domain that was in fact far out of his or her control.[35]

In a similar fashion, James Ryan argues that British photographs of India taken in the nineteenth century both created and entrenched the imperial project. Images understood to be "typical" allowed the distant British viewer to explore the world in a knowable, familiar manner. Furthermore, Ryan suggests that the style of these photographs frequently contained an implicit invitation to judge the subjects, whether they be landscapes or people. Photos of abundant yet uncultivated landscapes, he argues, allowed the viewer to assess the need for colonial intervention and improvement, much like the *Geographic*'s photographs of "Our Progress in Havana"—covered in chapter 3—demonstrated the benefits of an American presence in Cuba. To Ryan, the photographers of the British colonies chose to frame their human and natural subjects as tamed, reduced from a potentially hostile form in ways that "enraptured but never threatened."[36] Both Ryan and Rothenberg emphasize this avoidance of threatening, challenging, or disturbing images in favor of a more familiar narrative of how the world *ought* to appear.

Such an unspoken ideal led to a rather rigid and ahistorical view of culture, and is nicely illustrated by another photographic example. In 1907 the magazine printed a picture of a smiling young Japanese mother carrying two children on her back (fig 7.3). The accompanying article, "Women and Children of the East," was standard fare for the *Geographic*. It gave readers a chance to see the spectrum of cultural difference through the universalized practice of motherhood. Ten years later, the same photograph was used to illustrate not just the universality of motherhood, but also the particular character of the Japanese—in this case their stoicism. The caption of the photograph reveals much about the Society's understanding of culture:

> Stoicism is more than a tenet with the Japanese; it is almost a religion, and the mother of these babes, if the hand of death were laid upon them, could with calm fortitude relate her loss to a stranger

Figure 7.3 This photograph was used twice in the *Geographic*, in 1907 and 1917, to illustrate the stoicism of Japanese culture and the universal sentiment of motherhood. [Negative by Eliza R. Scidmore.]

without the display of grief, for it is a cardinal principle of her politeness that she should never burden another with her woes. But beneath this cross-barred cradle of cloth there beats the universal mother heart—universal in its high hopes for her children's future and in its eager joy at personal sacrifice for their happiness.

That the photograph was used twice suggests two important points. First, it gives us an example of an image that—by virtue of its popularity with the editors—speaks directly to the ideals of the magazine. In this case, the photograph captured both stoicism and sentimentality, which made it precisely the type of photograph most popular with the *Geographic*, one that would allow the viewer to empathize with the subject to a limited extent but to see difference as well.[37] Second, the repeated use of the Japanese mother demonstrates what Tamar Rothenberg has described as the *Geographic*'s implicit suggestion of "place-bound cultural timelessness." Here it made little difference that the photograph was dated—by at least a decade—for the caption suggested the enduring and unchanging nature of Japanese culture. This is particularly significant given that Grosvenor insisted upon "absolute accuracy" and claimed to reject photographs that were in any respect outdated.[38]

WORLD WAR I AND THE POLITICS OF NEUTRALITY

In order to keep up on the "geography of current events," Bell asked Grosvenor in 1907 whether he was prepared for the impending death of Emperor Franz Joseph. Bell anticipated the breakup of the Austrian Empire, and an ensuing scramble for the remains of the empire, and in light of this advised Grosvenor to collect anything he could find related to Austria. If his plan worked, Bell speculated, "the magazine would gain great éclat and be used by the press of the country as the great reservoir of facts from which they can draw."[39] Here again is the perception that the *Geographic* was not just another magazine, but a quasi-official organ of foreign policy that counted among its contributors Taft, Roosevelt, and Gifford Pinchot. This perception was confirmed by the Society's sponsorship of Peary's expedition to the North Pole, and then by its vigorous defense of Peary's claims against Cook's. A few years later, the Society solidified this reputation by sponsoring, along with Yale University, Hiram Bingham's "discovery" of Machu Picchu. Bingham's initial trip had been completed in 1912, and the results published in the *Geographic*, so sponsoring his return was precisely the kind of venture the Society sought.[40] Through expeditions to the Andes and the Poles, the Society would discover new worlds for its members, who themselves contributed through their membership fees. This mutual relationship was crucial to the early success of the magazine.

By the eve of the war Grosvenor had the confidence (in hindsight, well-founded) to enlarge his vision for the *Geographic*. Former Society president Henry Gannett excitedly reported that the *Geographic* had been ubiquitous in his travels through Mexico in 1912.[41] The Society also began to receive more letters and materials from members and potential contributors, and by 1909 Grosvenor was receiving one hundred times more photographs than he could ever use. This abundance of visual material allowed the editors to be increasingly selective, and to cultivate the *Geographic*'s reputation as a lavishly illustrated monthly that occupied the space between scientific journals and middlebrow mass-market magazines.[42]

1914 brought an explosion in readership and a fair degree of public attention to the *Geographic*, and introduced an era of political conflict that supplied endless material for the "geography of current events." Throughout the year the magazine paid close attention to Mexico and Europe, and Bell urged Grosvenor to gear the *Geographic* to these and other trouble spots around the world. As Bell wrote in 1914,

> I think that one of the historical features of the present struggle that will be permanently interesting to the world would be a description of Germany's attempt to undermine Great Britain's sea power. It might pay to have an enlarged map of the Kiel Canal and Germany's coast defenses on the North Sea. . . . It is the land battles that the newspapers and magazines most enlarge upon, and the public will be sick of these by the time the *Geographic* comes out. A broad and comprehensive treatment of sea power should, I am inclined to think, form the central point in the magazine.[43]

As usual, Bell's instincts were right. The intrigue of foreign affairs proved highly profitable to the *Geographic*, and by the end of 1914 membership had increased by 100,000—its largest annual increase to date. With membership now at 337,000, newspapers and magazines took notice, one repeating the Society's own statistic that "there is not a hamlet of 100 white population in the whole United States that does not have a member of the National Geographic Society among them."[44]

If Bell is credited with the impulse to popularize the *Geographic*, Grosvenor should be acknowledged for his attempt, however flawed, to incorporate a positive vision into the magazine during the 1910s. The remarkable growth of the Society in the decade after 1905 inspired Grosvenor to articulate the convictions that would guide his vision for the *Geographic*. Unsurprising were the first three: that the magazine would print only accurate facts, beautiful illustrations, and articles of permanent value. Yet the next four goals suggested just how complicated the *Geographic* had become:

4. All personalities and notes of a trivial character are avoided.
5. Nothing of a partisan or controversial character is printed.
6. Only what is of a kindly nature is printed about any country or people, everything unpleasant or unduly critical being avoided.
7. The contents of each number is [*sic*] planned with a view of being timely. Whenever any part of the world becomes prominent in public interest, by reason of war, earthquake, volcanic eruption, etc., the members of the National Geographic Society have come to know that in the next issue of their Magazine they will obtain the latest geographic, historical, and economic information about that region, presented in an

> interesting and absolutely non-partisan manner, and accom-
> panied by photographs which in number and excellence can
> be equaled by no other publication.[45]

Philip Pauly has taken these principles to be honest measures of the *Geo-graphic*'s content, observing that "everything unpleasant about a situation was left out, and if a group of people were controversial, they were not featured."[46] But the Society's principles of sunny neutrality are irreconcilable with the actual content of the magazine. The strong positions taken during World War I were just the more obvious examples of the magazine's more pervasive pattern of viewing the world through the lens of American values.

Consider the August 1916 piece touting "What our country is doing for Santo Domingo, Nicaragua, and Haiti." Written in the spirit of 1898, the article's endorsement of American armed intervention in these small countries made no pretense of honoring the seven principles articulated just a year before. With a people who "care less for the morrow than for the pleasures of the moment," it was no wonder that Americans were so often forced to intervene.

> Dominicans, realizing that the Monroe Doctrine is determined to
> afford them protection from their own excesses, their own bitter
> passions and blind purposes, have accepted the inevitable and have
> secured the blessings of peace from without when they could not
> attain that end themselves. . . . Unless the United States was to be
> forced to abandon the Monroe Doctrine, it had either to deprive
> other countries of their remedies or else intervene itself.[47]

Such a passage conformed to none of Grosvenor's principles. By any measure the *Geographic* failed to meet these standards, not just in overtly partisan articles such as those describing the "degeneracy" of Haiti, but in the very premise of the principles themselves. To have a magazine of human events that was timely and accurate, but that shunned all things critical, was an impossible idea.

Though violated in some way each month, the principles were significant for what they reveal about the magazine's public image. First, the very decision to create these principles indicated that change had come to the Society. With a magazine that covered the "world itself and all it holds," what might pass as unproblematic to the narrow, homogeneous readership of 1898 could stir widespread protest in 1917, when at least some of the half million members were bound to find fault with the Anglophilic perspective that pervaded the magazine. Thus the principles reflect a heightened awareness among the editors that theirs was a qualitatively more diverse audience that demanded a more careful, prudent approach. Even more important, the principles suggest that Grosvenor believed the magazine could occupy a space above politics and controversy, a particularly attractive proposition at a time when such tremendous political shifts—immigration, labor unrest, political conflict at home and abroad—beset the nation. In a world of upheaval, the *Geographic* would rise above the muck of politics and espouse a neutral—quintessentially American—position. In 1914 nothing could have been more elusive.

This attempt to stand above politics is what Rothenberg refers to as the *Geographic*'s "strategy of innocence," a phenomenon rooted in Grosvenor's assertion that American values were themselves universal. In the same issue that announced the magazine's principles—March 1915—the *Geographic* featured a seventy-page photographic celebration of the architectural landscape of the national capital, narrated by former president Taft.[48] The *Geographic* used this feature to frame Washington, D.C. as both American and cosmopolitan, existing almost outside of history. This sentiment surfaced again in a February 1917 photographic essay titled "Little Citizens of the World." Here the *Geographic* explored cultural difference through photographs of children at play in a range of international settings. On one level the essay seemed to celebrate the universal experience of play when viewed through the eyes of a child. Yet the *Geographic*'s own position was betrayed toward the end of the essay. The penultimate photograph of the series, of three young white girls skipping along a towpath, was captioned "just plain Americans with no strange customs" (fig. 7.4). Americans, in this view, stood above cultural particularism to embody a higher culture, demonstrating visually Pauly's general observation that, in the pages of the *Geographic*, "America was the future; the rest of the world the past."[49]

This sentiment—reflecting both exceptionalism and universalism—could of course have progressive implications. For instance, the Society adamantly criticized Wilson's new literacy requirement for immigrants through a long photographic essay in the February 1917 issue of the *Geographic* celebrating the contributions of recent immigrants to the nation's moral, military, and educational strength.[50] Five or ten years earlier, the Society would have endorsed Wilson's decision to weed out illiterates as a rational and self-protective measure against the "hordes" arriving daily at Ellis Island. But on the eve of a war it was eager to join, the Society was devoted to accelerating the Americanization of immigrants and bolstering nationalist sentiment. It considered the moment one of international unity, when immigrants coming from all parts of Europe would aid in the fight against a universally acknowledged enemy. In this context to encourage immigration was a national necessity, just as the insistence on immigration restriction had been a decade earlier. While in 1907 the magazine warned of the dangers of unrestricted immigration, in 1917 it found room in the nation for an additional ninety million citizens. Even in its defense of immigrants, the *Geographic* had violated its own principle of avoiding anything "of a partisan or controversial character."

By early 1917 the magazine was almost entirely devoted to the war, making any pretense of neutrality as moot as it had been in 1898. Apparently innocent profiles of the orphans of Belgium and the importance of British seaports thinly disguised the Society's sympathy for the Allies. The clearest—and possibly most consequential—failure to measure up to the standards of neutrality came in the March 1917 issue. Before Congress had declared war on the Allies, the *Geographic* carried two articles invoking the nation's "natural" ties and commitments to the British. Sydney Brooks, a British national, celebrated British citizens as the unsung heroes of the war, and lamented their tendency to exaggerate their failures, downplay their victories, and hide their sacrifices. Brooks took it upon himself to correct the "misinformation" left by the excessive modesty of the British, and exalted the "intense determination" of the 450 million loyal subjects to win

Figure 7.4   One in a series of photographs profiling the wide range of customs of children around the world. This photograph's caption, "just plain Americans with no strange customs," reflects the *Geographic's* sense that Americans were, in a sense, free of culture-bound traditions. Gilbert Grosvenor/NGS Image Collection.

the moral battle raging across Europe. He closed with a flourishing call to the Allied cause that shocked more than a few of the Society's members:

> from this welter of renewal there is springing up an England strengthened by enormous sacrifices for great ideals, ennobled by poverty, disciplined without losing her characteristic flexibility and self reliance, knowing more than a little of the true faith of social equality, and proud to have played once more, and not without honor, her historic role as the defender of the liberties of Europe.[51]

U.S. Senator John Sharp Williams enlarged on Brooks's themes in an address to the Senate on April 4, revised for publication in that same issue of the *Geographic*. Reproaching the nation's partial sympathy with Prussia and dismissal of Britain's plight, Williams challenged his readers: "Which would you rather do—fight Prussia now, with France and England and Russia to help you, or fight her later, when she is foot-loose, by ourselves? You have got to do one or the other." In contrast to the freedom-loving British he placed a Prussian "race infected with the poisonous idea that it is ruling by divine ordinance."[52]

The articles by Williams and Brooks were objectionable for their assumption of an Anglophilic spirit among their readers that no longer existed, especially prior to an American declaration of war. Here Grosvenor must have realized just how much had changed since 1898. Similar articles endorsing the nation's posture in the Spanish-American War caused no uproar from the *Geographic*'s small, homogenous readership. But twenty years later that homogeneity had vanished, replaced by a rash of requests to cancel memberships. One member angrily responded that "I wish no kinship with Sydney Brooks and John Sharp Williams." Hostile letters chastened the editors for printing such "bombastic slush" and for "stirring up race hatred." Typical was the sentiment of Charles Todd of St. Louis, who asked that his membership be terminated because "the March issue is devoted mainly to the unqualified land of the greatest of robber nations, England, to deceive the American people." In a barb that must have been particularly painful for Grosvenor, another member concluded that the *Geographic* "has sunk to the low place of a common political Magazine" in violating the principles Grosvenor himself had established in 1915. An anonymous German newspaper editor concurred, concluding that with the March issue the *Geographic* had "lowered the flag of scientific impartiality."[53]

After the nation's declaration of war in April, the *Geographic* quickly sympathized with the Belgians and vilified the Germans in order to demonstrate its loyalty, and in all printed fifty-seven articles related to the war. Still, the magazine's belligerent posture angered the Society's members. In June 1918 Secretary of State Robert Lansing's "Prussianism" article drew fire from members for characterizing the German system as "sinister" and the Germans as "huns."[54] One Midwesterner complained that too "much space is devoted in the magazine to British propaganda, and is as bad as the propaganda the Kaiser spread in this country before and during the war."[55] Particularly offensive to this reader was an incident related in an article that described life among Armenian refugees in the aftermath of war. The American author of the article exalted the work of the British soldiers as representative of their high degree of civilization. When one British soldier in particular was asked about the destination of the starving Armenian refugees, he replied callously that "they sleep and die on station platforms or else in the trains; and so it goes, week after week." This comment passed into the article as reportage, not commentary. The implication was clear: if British soldiers represented the highest stage of civilization, their Armenian charges had not yet passed through the earlier stage of barbarism.[56] Many members had no easy time squaring this editorializing with the principles Grosvenor had articulated for the *Geographic*.

These comments suggest that some members had in fact taken the magazine's pretense of neutrality at face value. One reader, for instance, was shocked to find that "we have to get Entente propaganda crowded down our throats even in the *National Geographic Magazine*."[57] The Society's reputation as a scientific and educational organization goes a long way toward explaining the widespread disappointment and outrage expressed by so many during the war. Others were troubled less by the position taken by the *Geographic* than by the decision to cover political conflict at all. A Minnesota pastor terminated his membership because the magazine carried "too much militarism to suit me." Another put it more succinctly: "I'm sick of war pictures."[58] All these comments indicate that, for all its limitations, the *Geographic* had carried an aura of respectability among its members that had been called into question by its actions during the war.

THE POLITICS OF RESPECTABILITY

In 1919, the Society endorsed the League of Nations through a feature by Grosvenor's cousin, William Howard Taft, even though Grosvenor himself was deeply ambivalent about Wilson's principle of self-determination.[59] In the same year Grosvenor celebrated twenty years with the Society, and was rewarded a year later by his election to the Society's presidency.[60] In that year he changed the bylaws in order to limit the power held by members residing in Washington, D.C., an especially important adjustment because to him a central feature of the Society's lure was the dimension of individual ownership.

> The fact that our plumber friend, our newsboy, and our millionaire who likes a good bargain, like The Geographic Magazine, is an equal proprietor and patron of these expeditions, inspires each with pride in his own participation, and he follows their progress, a zest and educational value to himself quite different from the feeling he would have if he had no share in the work; it is his Magazine, his expedition, which makes all the difference between his baby and somebody else's baby.[61]

Grosvenor had become a minor celebrity in Washington, the leader of a quasi-official organization worth millions.[62] The father of American academic geography, William Morris Davis—who in 1900 had tried to oust Grosvenor as the assistant editor of the *Geographic*—hailed the Society as a "national asset" in 1924.[63] The Society's general membership stood at nearly 900,000, and the *Geographic* had already become standard fare in dentists' offices and school libraries. The Elmhurst (Illinois) *Press* reported that 75 percent of the town's homes took the *Geographic* in 1926.[64] When queried about the *Geographic*'s success, Grosvenor repeatedly pointed to the seven principles of 1915, yet his assessment is problematic in two regards.[65] First, despite the overall growth, the years from 1915 to 1920 had been a time of stagnation for the Society. The readership's dissatisfaction with the *Geographic*'s coverage of the war, along with rising dues and a general economic dip, all curbed the Society's growth during and just after the war. Second, if the war had proven anything to the Society, it was the logical impossibility of a current events magazine that transcended politics. And yet, even if we understand the flaws in Grosvenor's vision, we must also acknowledge that it was shared by millions. Consider the following verse, sent by a member to the Society in 1921.

> The Geographic comes to me,
> Filled with scenes from 'cross the sea,
> Pictured love of distant lands;
> Salvaged drift from foreign strands;
> A flag-decked car in Kyoto's streets;
> A moon-lit "banker" with ghostly sheets,
> Circled with smoke in my easy chair,
> I'm a boy again, and rise in air,
> On the back of Hindu's wooden horse,

And, turning the pages, I steer my course
to all the places I long to be;—
When the Geographic comes to me!

When the Geographic comes I go
On fairy trips to Mount Nikko,
Whose lacquered temples, roofed with gold,
The shelt'ring cedar forests hold;
Or, 'midst the varicolored blooms
Of iris, where great Fugi looms.
Above Lake Shoji's placid face,
I wander, by the magic grace
Of Walter Weston's graflex art.—
Though safe at home, I form a part
Of storm-tossed crews on Arctic sea,
When the Geographic comes to me.

Burma, Bombay, Baluchistan,
"Shooting Rapids in Japan"
Hopi maids in the "altogether,"
French m'selles in Breton heather,
"Sixteen knots in December blow"
Ama-No-Hashidate covered with snow,
"The Crown Lady's Tomb" at sunset time,
The "movie" sheets of a Nippon mime;—
All these wonders, and more, are mine,
From the Devil Dance in a Lama shrine,
To "Ups and Downs in Brittany,"
When the Geographic comes to me.[66]

However clumsy, the verse suggests that our own image of the *Geographic* has a long history. Despite its wartime stumbling, the *Geographic* had by 1921 come to symbolize something that endures to this day.

In fact the *Geographic* took on its modern form and style in these years, coming to rely primarily on photography for its success. Photography translated distant lands and complicated scientific phenomena into easily discoverable realities. And among the most memorable of these photographs for white Americans were those that combined racial and sexual taboos— especially those of African women. These were the photographs that elicited strong responses during the 1920s, in part because the armistice allowed the *Geographic* to resume coverage of areas that had been somewhat neglected in recent years, particularly those outside Europe and North

America. This was, however, overwhelming to some readers. One thanked the Society for its March 1925 issue, which focused on "white people in Europe. Keep up the good work in this field of geography. Your readers are all quite tired of Negroes and scenes in Africa."[67] In fact, from 1924 through 1926 the Geographic carried only eight articles on Africa, relatively few considering the magazine printed more than 120 articles in total during these years, and devoted about the same amount of space to Asia and South America. Yet this coverage of Africa was still considered excessive by a few of the Society's members, particularly when it challenged ideas of propriety. A reader from Ontario wrote that "we are fed up with these black naked people that appear in every number."[68] Another agreed, deciding that the Geographic contained "too many naked savages. Do not object to nakedness but to quantity."[69] A New Yorker was similarly put off by the Geographic's long feature on an automobile journey through Africa, because "looking at pictures of people from the jungle that don't pay much attention to clothes don't interest me at all."[70] These photographs clearly left strong impressions in people's minds. Though perhaps with a greater degree of self-awareness, modern poet Elizabeth Bishop recalled a 1918 issue of the Geographic, showing "black, naked women with necks wound round and round with wire, like the necks of light bulbs. Their breasts were horrifying."[71]

And yet despite its relatively provocative photography, the Geographic continued to stand largely beyond reproach in American culture. Nudity, however common and offensive to some, did not ever inspire widespread objections, which suggests that the Geographic was able to legitimate its work as "scientific" to the one million amateur anthropologists who received the magazine by 1926. While some were repulsed by the presence of native Africans and Indians in the magazine, others, such as this New Jersey pastor, had precisely the opposite sentiment.

> Allow me to say that I am acquainted with your wonderful magazine and am very grateful for the opportunity of receiving it. You are supplimenting [sic] the work of our various educational institutions in a most remarkable manner in sending this magazine into the homes of our people who are to become the backbone of the nation. In my pastoral visitation I frequently find copies of the magazine in homes and what will delight you to hear from a minister, very very often in the hands of promising young people, whom I have observed in a reading position when ushered into many homes. I have had parents tell me how they appreciate your magazine; I have had the young people themselves tell me how they

admired it. In a day when much of the fiction which comes from
the press is not fit for the growing minds of our young folks to be
nurtured upon it is indeed gratifying to know that there are many
who will be the leaders of tomorrow who are absorbing something
from the pages of your periodical which is always helpful in illumi-
nating the mind.[72]

This comment reveals the *Geographic*'s uncanny ability to balance its potentially contradictory roles: guardian of propriety, scientific authority, and
entertaining diversion. The Society's careful negotiation of such wide-
ranging aims, though not without periodic missteps and failures, does much
to explain the astounding success of the *Geographic*.

In a 1929 letter to one of the magazine's contributors, Grosvenor wrote
that "one of the reasons that The *Geographic* is so widely read is that our
members know that they can turn in confidence to The Magazine for interesting, pleasant information on geographic subjects; that they can find
in it *mental relaxation without emotional stimulus*."[73] Ironically, though, this
policy of avoiding "sensational, alarming articles" led the *Geographic* to
condone one of the most alarming developments of the twentieth century:
the rise of fascism. The insistence—however unrealistic—that politics be
avoided at any cost, that only the bright side of life be exposed, was brought
to its logically absurd conclusion when the *Geographic* covered the "modern" changes taking place in Berlin and the "rebirth" of Rome. In stories on
Nazi Germany and Fascist Italy—as well as Fascist Spain—the *Geographic*
exalted the "progressive" spirit of change in these countries, without any
kind of evaluation of their costs. Paging through these abundantly illustrated issues of the mid-1930s is a chilling experience. Bright color photographs of Berlin—teeming with Hitler's thronging supporters—visually
celebrated the rise of Nazism as a sign of recovery in long-suffering Germany. Public squares, bedecked with swastikas, were narrated as symbols of
national pride and a forward-looking citizenry. Fascist youth organizations
in Italy and Germany were portrayed as popular clubs for enthusiastic boys
eager to absorb their national traditions, while military displays were jaun-
tily described as evidence of civic-mindedness and national independence
(figs. 7.5 and 7.6). When Douglas Chandler, author of the *Geographic*'s feature article on Germany, "Changing Berlin," was discovered to have been a
German spy, the Society managed to escape relatively unscathed.[74]

Surely we cannot hold the *Geographic* accountable for the cataclysms
that few others anticipated with the rise of fascism, but given Grosvenor's
own pretense of transcending politics, how could such articles have passed

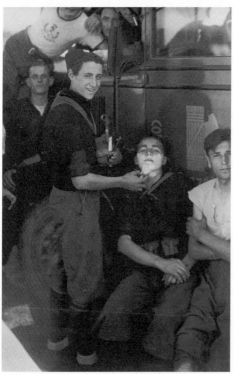

Figures 7.5 and 7.6 *Above,* One of
the many neutral, if not positive,
illustrations of life under fascism
in the *Geographic,* this one from
"Changing Berlin" in 1937. The
photograph's original caption reads:
"Banners over Berlin—A Bright,
Sunshiny Day, with *Unter Den
Linden* in Gala Dress." Wilhelm
Tolbien/NGS Image Collection.
Left, Another happy view of life
under fascism in the *Geographic,* this
from "Imperial Rome Reborn" in
1937. The original caption reads:
"Shave Me and I'll Shave You—
A smiling Young Fascist operates on
a comrade in camp, while another
bristly face waits its turn." On the
truck door is the symbol of Italian
fascism. Bernard F. Rogers, Jr./NGS
Image Collection.

into the magazine without extensive editing? As we have seen, the *Geographic*'s claim to neutrality thinly veiled its own politics. But this *ideal* of neutrality—though compromised—had tremendous appeal in a nation that had yet to determine its role abroad. Thus the coverage of fascism was brought in through the back door: Grosvenor, like many other Americans in the 1930s, accepted fascism as a bulwark against communism.[75]

The *Geographic*'s reluctance to engage explicitly in politics was designed to deliver "mental relaxation without emotional stimulus." If, as Richard Ohmann argues, the magazines of the 1890s helped Americans feel at home in a new era of consumer capitalism, the *Geographic* made Americans feel at home in the world by transforming controversy into wholesome fare for popular consumption. As one New Yorker commented, the magazine "is of special benefit to stay-at-home folks but should one happen to visit strange lands I think it would seem more like re-visiting after reading the wonderful descriptions in the *Geographic* and looking at the pictures."[76] By drawing on familiar narratives, cultural stereotypes, and persuasive images, the *Geographic* affirmed members' understanding of the world—just as it had with their understanding of science—and gradually expanded it. In this regard we must certainly credit Grosvenor's vision—however flawed—as one that appealed to the American public by striking a careful balance between the foreign and the familiar. Toward the end of his tenure at the Society, Grosvenor proudly recalled that, in the opinion of some professional geographers, he had "vulgarized geography." It was a curious recollection, for throughout his life Grosvenor had claimed precisely the opposite, that it was his "seven principles" of propriety and neutrality that ensured the *Geographic*'s success. The apparent contradiction is not insignificant.[77]

8

The Map and the Territory

1900–1939

The Spanish-American War, along with the Open Door Notes and the Russo-Japanese War, sparked in Americans a keen interest in geography. Cheap monthlies and newspapers were full of cartoons that used cartographic imagery to persuade readers of America's opportunity and mission abroad (figs. 8.1 and 8.2). With illustrations of Uncle Sam extending his reach over the Pacific or of Spain's retreat across the Atlantic, the popular press translated the distant geography of the war into a conquerable spatial narrative that endorsed the nation's leap around the globe. Similarly, while the atlases of the 1870s and 1880s stressed the gulf that distanced Americans from the world, those of the early twentieth century presented a more integrated global community that centered on the United States.

But while the pace of foreign activity in the 1890s recast many conventions in the world atlas, the style and appearance of the maps themselves emerged relatively unchanged. In fact, early twentieth-century cartographic companies actively avoided changes in their maps and atlases in order to meet what they perceived to be public expectations. These decisions had important implications for American cartographic culture, and for the public's understanding of the relationship between the map and the territory. Atlases and maps also attracted a wider audience in the early twentieth century, promoted by conflicts abroad and a growing leisure market at home, both of which encouraged map firms to adopt more aggressive and sophisticated strategies in the hope of controlling an increasingly national and competitive market. How mapmakers negotiated this growth—by both accepting and resisting change—is the subject of this chapter.

IT OUGHT TO BE A HAPPY NEW YEAR.
Uncle Sam and his English cousin have the world between them.

Figure 8.1 Imagined geography: cartoon from the cover of *Judge* magazine, January 1899

UNCLE SAM'S STRING OF CANNON CRACKERS.
July 4.

Figure 8.2 Imagined geography: cartoon from the *Minneapolis Journal* (1898)

THE ENTRENCHMENT OF AN AMERICAN STYLE

Rand McNally and Cram were joined at the turn of the century by a new rival, based not in Chicago but in New York. Caleb Hammond spent twenty years as a drafter at Rand McNally before establishing his own map company in 1900, and as a result the production, distribution, and appearance of his maps were strikingly similar to those of the other major map companies. As in the past, many contemporary cartographers and geographers considered these mass-produced atlases aesthetically inferior to their European counterparts, and assumed that this was an inevitable consequence of American production techniques. Instead of limiting the number of place names to emphasize the largest or most important, American firms loaded the maps with as much information as possible. This practice stemmed in part from the fact that the United States was producing cartographers with little or no generalization experience, trained instead by an industry where draftsmen were rapidly becoming obsolete. With the need substantially diminished, fewer and fewer were trained with these skills and the prevailing style began to entrench itself, transforming a historical practice into an American cartographic ideal, an accidental aesthetic that transcended the circumstances of time and technology.[1]

Despite the inflexibility of the maps themselves, the early twentieth century represents a turning point for the American world atlas generally. In the 1880s and early 1890s, from 75 to 80 percent of the atlas maps were devoted to coverage of the United States, but after 1898 this figure declined to about 50 percent, which made the twentieth-century atlases significantly more international than their nineteenth-century counterparts.[2] As late as World War I, Rand McNally's *Imperial Atlas*—the name itself significant— opened with a map of America's epic growth across the West and around the world (see fig. 2.7). The *Imperial Atlas* had reconfigured the nation by enlarging its borders beyond the continent. As one reviewer commented, "it does look a little bit odd to see Porto Rico, Hawaii, and the distant Philippine islands on the United States map. But they are there and printed as carefully and described as carefully as if they had been for a whole generation in their present honored company."[3] As if to confirm this as a turning point in American history, Rand McNally grouped the maps of these island territories alongside those of the United States, despite the geographical gulfs that separated them. By contrast, maps of the European powers were separated from those of their own colonies in contemporary atlases, a cartographic rendition of American exceptionalism.

The Cram Company also emphasized America's recent growth by opening their *Ideal Reference Atlas* of 1902 with large maps of the nation's territories and of the proposed canals in Nicaragua and Panama, while leaving all of South America to be covered in just two maps. Rand McNally's *Imperial Atlas* of 1904 boasted four maps of Manila Bay but only three for all of Africa; its *Library Atlas* of 1912 introduced an elaborate map of the Philippines, while one marketed in 1915 devoted two full pages to the West Indies. The detail of these maps—like American interest—peaked prior to World War I.[4] As one Rand McNally employee candidly remarked, the degree of attention given to any region was a function of, among other things, its "relative commercial or industrial importance," and Hammond's *Pictorial Atlas* of 1912 illustrates this principle with its map of the world, organized not according to languages spoken, but according to the languages used to conduct commerce.[5]

The changing narrative of the atlas over time also indicates the fluid nature of the genre. The 1887 edition of the *Pocket Atlas*—Rand McNally's most popular series—described Japan as a social hierarchy with absolute monarchical government and compulsory school attendance, then briefly enumerated its chief agricultural products. Yet by 1900 the atlas focused on Japan's rising manufacturing, trade, and mineral wealth, and described principal cities in terms of such commercial infrastructures as the extent of rail connections, ports, and industry. The Philippines, briefly passed over in the 1887 edition, were lovingly described by 1900 as having both "undeveloped" and "unsurpassed" resources.[6] And compare the profiles of Cuba in 1887 and 1900: in 1887 the colony was profiled briefly in a larger section on the West Indies through statistics on population, ethnic breakdown, geographical features, mineral wealth, and education. Not surprisingly, in 1900 the protectorate was given its own section apart from the West Indies, with a more comprehensive history of the island, including dates of discovery, exploration, wars, and emancipation, and the details of the American occupation. Following the history were extensive descriptions of Cuba's climate, forests, and mineral wealth, accompanied by a substantially more hopeful profile of its resources.[7]

> Forests among most valuable resources of island. . . . Soil of almost inexhaustible fertility and highly favorable climatic conditions entitle Cuba to rank among the foremost agricultural countries of the world. Resources, however, largely undeveloped, but possibilities of the island are almost incalculable. . . . Minerals abundant and valuable.

With "innumerable varieties" of fruit trees that grew "luxuriantly," the atlas elaborately anticipated the island's trade potential. All these qualities gave Cuba tremendous commercial promise, a far cry from the description of 1887. Yet while in the 1900 edition the atlas pronounced Cuba's mineral wealth "abundant and valuable . . . in some, deposits are inexhaustible," by 1936 enthusiasm had died and the minerals were simply noted as "not of great commercial importance."[8] In these and other instances, characterizations of the natural world were themselves negotiable, as subject to change as political boundaries or foreign policy.

This growing interest in the world as an arena of commercial activity occasionally challenged the deterministic frameworks that had reigned in late-century atlases. In the 1880s and 1890s, climate and race served to homogenize and isolate certain geographic areas and to place them within a rigid hierarchy of civilization. By contrast, the new interest in the realm of commerce allowed for more differentiation between nations. For example, no longer were the Philippines just islands of the Orient with a climate ill-suited to the white race. After the islands became the direct responsibility of—and a potential economic asset to—the United States, they became substantially more complex. Though race still operated to classify the Filipinos as "our little brown brothers," the focus on commerce made the islands less foreign to Americans. Commercial geography had a democratizing influence on the atlases; even though racial hierarchies persisted, they now competed with commercial ones. No longer would countries and territories be described solely—or even primarily—according to race and climate; the ability of their inhabitants to extract, transform, and trade natural resources had become perhaps their most salient feature.

WORLD WAR I

Europe's descent into war generated a slew of popular war atlases in America, almost all of which were produced by private map companies. In many cases, though, the firms had little incentive to design maps specifically for the conflict, and instead simply repackaged existing maps of the war-torn regions. Symbolic was Rand McNally's *Atlas of the European Conflict* (1914), which opened with a map of the world centered on the United States, an orientation that necessitated dividing the arena of conflict in Europe.[9] In the *Graphic Representation of the Battle Fields of Today*, Rand McNally used an existing commercial map of Germany that featured towns, sea routes, and shipping schedules. The map was difficult to read, overlettered, and strewn with details appropriate to an interest in peacetime commerce rather than wartime strategy.[10] Much of the difference between the European- and American-made war maps was a matter of scale: generally the former depicted areas on a larger scale that allowed for clearer relational descriptions.

But because the American maps were taken from existing collections their scales were qualitatively smaller, unsuitable for more than a cursory over-view of the battlefields.[11]

Even the National Geographic Society, highly sensitive to public taste, replicated existing cartographic styles. The Society had long issued maps with its magazines, the earliest of which were designed to chronicle polit-ical conflicts such as the Sino-Japanese War, the Spanish-American War, the Boer War, and the Russo-Japanese War. But it was not until World War I that any of these were actually created by the Society. The 1918 map of the western front—the first to be drawn by the Society's new Cartographic Division—looked much like those made by Rand McNally, dull in appear-ance with the overall contours of battle lost in an infinite jumble of place names (fig. 8.3). Care has been taken to reproduce this map detail in its ex-act size in order to illustrate its general character; the level of locational de-tail apparent in this section is not unique, but is replicated throughout the map. Ironically, on this "war" map there is in fact little indication of the general contours of the conflict. Only faintly do rivers appear, and to the naked eye the map is difficult to read without magnification. Yet this emphasis on locational detail—even at the expense of other relation-ships—was not unintentional. The Society's goal was to bring out a map where "practically every name in the battle area, however unimportant, might have its place," where "fully 95 percent of the names mentioned in the daily news appear."[12] The Society actually boasted that this little map—measuring approximately 28 inches by 33 inches—squeezed in in-formation from French and German maps four times as large.

It was this apparent flaw—the inclusion of every conceivable place name regardless of its insignificance—that was eagerly welcomed by some of the Society's members. One who kept the map on his office wall was thrilled by its inclusion of "more towns and villages than any other I have been able to secure," for it was this quality that allowed him to fol-low the battles with precision.[13] Albert Holt Bumstead, the National Ge-ographic Society's chief cartographer from the 1910s until his death in 1940, echoed this sentiment in a letter to the Geographic's editor, Gilbert Grosvenor, in 1915. Discussing his planned improvements for the Society's maps, Bumstead decided to erase the contour lines that marked elevation, explaining that "contours mean much to me, but I must admit with disap-pointment that to most map users they are nothing but a confusion. Ele-vation is probably the least important of the information given on the map, so lets [sic] not sacrifice the clearness of anything else for its emphasis."[14] Caleb Hammond, head of the Hammond map company from 1948 to 1968, concurred: Americans made sense of maps through towns that were

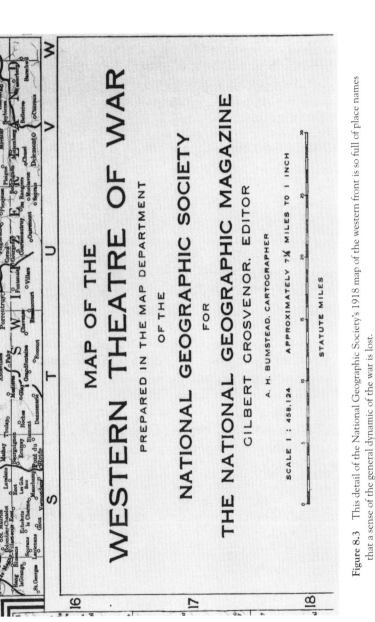

Figure 8.3 This detail of the National Geographic Society's 1918 map of the western front is so full of place names that a sense of the general dynamic of the war is lost.

relevant to them, and relief markings simply competed with and detracted from that goal. Occasionally the public confirmed this assumption, as in this suggestion sent to the National Geographic Society during the First World War: "I would like to have an *atlas* showing every TOWN in EUROPE and ASIA big enough to have a POST OFFICE and every STREAM long enough to have a NAME."[15] Maps were useful and valuable to the extent that they allowed Americans to identify multiple, discrete locations, even though it was precisely this feature that many European and American cartographers found so maddening. This interest in identifying locations rather than understanding the landscape as an entity continually characterized the American culture of cartography, perhaps a legacy of the gazetteer tradition of the early nineteenth century.

Though the maps in war atlases were slow to reflect change, the text quickly adapted to shifts in American foreign policy, as it had in 1898. From Archduke Franz Ferdinand's assassination in 1914 to Woodrow Wilson's declaration of war three years later, the United States maintained formal, if not actual, neutrality in Europe. Ethnic divisions and the desire to continue trading with members of both the entente and the alliance delayed what might have been an earlier commitment to defend Britain, and so the war atlases — as perfectly aligned with the state as they had been in 1898 — reflected the same. Rand McNally's 1914 *Atlas of the European Conflict* characterized the war as a result of "the thirst for aggrandizement of empire, political, military, and commercial, and the mutual fear and jealousy of kings." For years, the atlas claimed, the world lived in fear of the inevitable clash between the nations of Europe, "each shouldering immense burdens of armament, each straining to surpass the other in strength, and power to destroy. Engines of war have been perfected until man's ingenuity in the preparation of catastrophic elements has been exhausted."[16] But after the nation's entry into war in April 1917, the company placed blame squarely on America's new enemies. Prussia had "foisted itself upon the confederacy of German states as the dominant power, the seat of an hereditary autocracy, and the controller of the constitution and the armed forces of the newly created empire." The war could now be understood as a result of Prussia's expansionist drive, yet this by no means translated into a denunciation of imperialism. As the atlas explained,

> The power and wealth of a nation may be measured to a certain
> extent by the amount of territory she controls at home and abroad.
> Every square mile of territory is a source of revenue and mineral,
> agricultural, or manufactured products, offers a field for export and

commercial exploitation, and yields land and customs revenue for the state.[17]

The atlases immediately integrated Wilson's decision to enter the war by vilifying the ideology of America's new enemies: it was world domination, not economic expansion, that the atlases judged unacceptable.

THE POLITICAL AS NATURAL

World War I brought dramatic upheaval to the boundaries of Europe. As outlined in Wilson's Fourteen Points, the Austro-Hungarian and Ottoman Empires were dismantled to make way for Czechoslovakia, Yugoslavia, and a newly independent Poland. States in the Middle East also gained independence with the breakup of empire, leaving Turkey a small state, and placing Palestine, Jordan, and what is today Iraq under the British, and Lebanon and Syria under the French. Africa was similarly redistributed among the victors. In Europe and the Middle East the war had dramatized the flexible nature of geography and had left many Americans confused. Cartographic companies were quick to capitalize on this sense of epic geographic change. The romantic adventures of Lindbergh's flight across the Atlantic and Byrd's explorations of the South Pole also contributed to this swell of interest in geography. Radio news broadcasts encouraged listeners to follow events abroad with an atlas, and the immensely popular children's program *American School of the Air* became required listening in 200,000 classrooms over its eighteen-year run from 1930 to 1948. Educational and popular radio programs exposed young listeners to world events, exotic locales, and the feats of American explorers in the interwar era.

The American public also became increasingly accustomed to reading maps due to the automobile revolution of the 1920s. Oil companies began to give away road atlases at service stations, a practice that brought countless Americans into contact with maps on a daily basis and prompted one comment that "map reading is no longer the trying, difficult schoolroom task it used to be." Within this responsive set of circumstances after World War I, Rand McNally embarked on an aggressive campaign to protect its market share and increase its sales through advertising, public relations, and even tariffs. In the process, the company secured not just dominance of the atlas market but also a reputation among Americans as a cartographic authority.[18]

Despite the upheavals in Europe and the expanding consumer market at home, the world atlas emerged from the war with much of its prewar form and content intact. Rand McNally's revised *Ideal Atlas of the World* introduced even more focused information about the resources of the world through a candid discussion of America's need for markets and the commercial gains it had made in the war. Alongside the continuing commercial focus of the postwar atlases was a gradually increasing interest in the world as a physical manifestation. In the nineteenth century, physical surveys of North America had been carried out by the United States Geological Survey and the Army Corps of Engineers, and the same agencies had surveyed the American territories in the early twentieth century; in the 1920s and 1930s, private cartographers conducted surveys of the United States, Europe, Japan, and South America. These new surveys of the non-American world drew attention to the physical world, and new maps began to emphasize the physical layout of the land by charting climatic patterns, elevation, and sea currents. Yet the major American map companies generally *excluded* this information from their popular atlases, judging it appropriate for school students but inappropriate—uninteresting and irrelevant—for the general population. This is nicely illustrated by the arrival of Rand McNally's *Goode's School Atlas* in 1923. Introduced to fill the rising demand for atlases in the secondary schools, the atlas was conceived and executed by John Paul Goode, a professor of geography at the University of Chicago and, after 1900, the chief cartographic advisor to Rand McNally. The atlas was a significant departure for the company. In fact, its very existence served as an indictment of the American cartographic tradition that Rand McNally had itself helped to create.

Until this point, American maps had consistently portrayed the world on the Mercator projection. This sixteenth-century projection translates the globe onto a cylinder, which causes great distortion at the northern and southern latitudes and fails to repesent comparative land mass accurately (fig. 8.4). The projection's power derives from its practical accuracy as a sailing chart, but over time it began to be used for other purposes, and gradually became the most common—and eventually the only—way to map the world. Widespread recognition of the limits of the Mercator projection did not develop until the 1930s and 1940s. Thus Woodrow Wilson was far ahead of his time when in 1913 he asked Americans to turn toward the globe in order to realize that nearly all of South America lay east of North America.[19] The concept of projection was simply not yet part of American culture, and would not be until well after World War I. By comparison, the 1882 edition of Stieler's *Hand Atlas*— one of the most widely printed German atlases—displayed no less than nine different projections on its title page, and in doing so implicitly suggested the complex nature of cartography. Such a display would not have been culturally meaningful for the American public until the mid-twentieth century.[20]

This made the publication of *Goode's School Atlas* in 1923 provocative. Goode argued that the reflexive use of the Mercator projection in America had been damaging to the nation's understanding of geography generally and of maps specifically. The projection, he explained, was useful as a navigational tool, but nothing more. Particularly irritating was the projection's distortion of land area, and Goode proudly insisted that in his own atlas "every square inch in the map represents the same number of square miles of the earth's surface as any other square inch in the map.[21] *Goode's Atlas* was an argument against the presumption that any single projection could accurately map the earth in all respects. Upon opening the atlas students were confronted with projections of all kinds that reconfigured and distorted the earth in startling ways, the "evil" Mercator projection most egregiously. And as students turned the pages of the new atlas they found a strange new alternative based on Goode's own homolosine projection, an attempt to correct for Mercator's long-accepted flaws (fig. 8.5).[22] With interruptions at the northern and southern latitudes, the new map was disorienting and challenged the existing cartographic culture. One employee called it "a very confusing book" for its depiction of the world as "four irregular ovals connected at the North Pole, one a bit longer than the others and with a jagged tooth on its eastern side." Andrew McNally II—then the company's president—recalled that although *Goode's School Atlas* sold well in schools, the unfamiliar homolosine world map made it insufficiently "unified" to pass muster with the general public.[23]

The orientation of Goode's new world map was equally jarring, for it was one of the first to move the United States off center. In this respect Goode challenged a longstanding tradition that dated back to the mid-nineteenth century, when the first map was published that placed the Western Hemisphere at the middle rather than on the left.[24] With few exceptions, atlases had divided Eurasia rather than sacrifice the centrality of the United States, a practice encouraged by the growing importance of the Pacific to the nation in the twentieth century. In fact, the National Geographic Society mapped the world with the United States at the center— almost without exception—until 1975.[25] This made the appearance of Goode's homolosine world map even more disruptive to a nation reared on the Mercator projection.

The content of *Goode's* maps was as disorienting as their shape. The first edition of the atlas did not even include a political map that divided the world along national lines. Though the later editions introduced more

Figure 8.4 The sixteenth-century Mercator projection

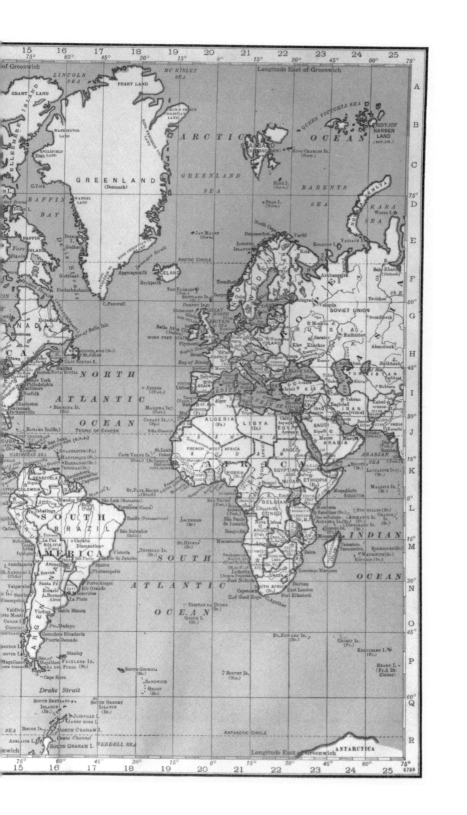

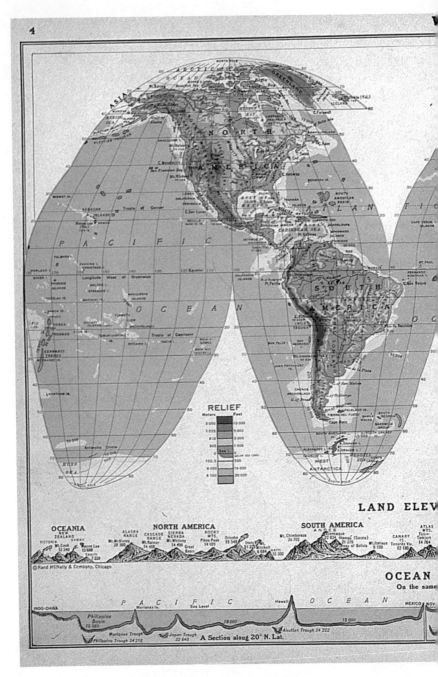

Figure 8.5 The twentieth-century homolosine (Goode's) projection

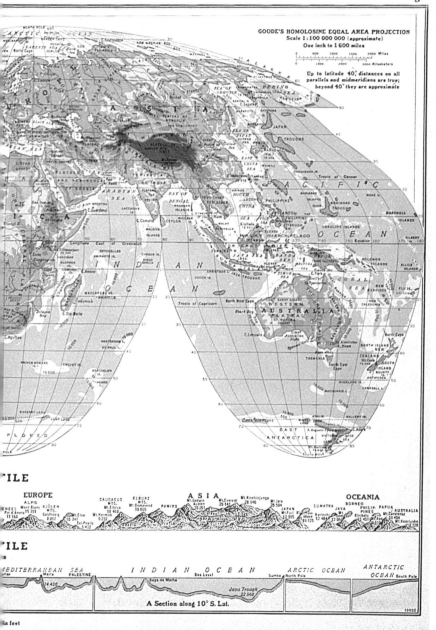

GOODE'S HOMOLOSINE EQUAL AREA PROJECTION
Scale 1 : 100 000 000 (approximate)
One inch to 1 600 miles

Up to latitude 40° distances on all
parallels and midmeridians are true;
beyond 40° they are approximate

A Section along 10° S. Lat.

in feet

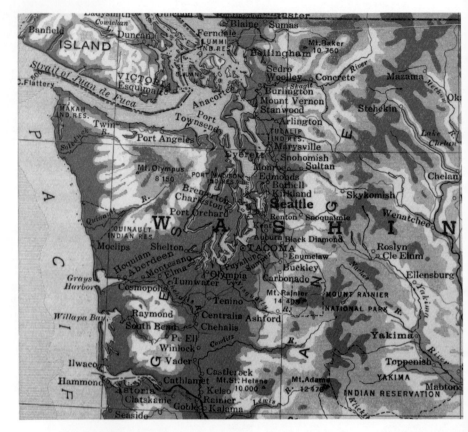

Figure 8.6   Notice the inclusion of relief on this detail of *Goode's* map of the Pacific Northwest. Compare with figure 8.7.

traditional political maps, far more central were the extensive physical and commercial maps that depicted ocean currents, climate, vegetation, elevation, soil regions, resource distribution, and trade. In fact, the first edition of *Goode's Atlas* omitted what had become the *sine qua non* of the atlas, a political map of the world that identified nations and colonial possessions. Instead, Goode emphasized the physical and commercial dimensions of world geography; not until later editions did he introduce political and territorial maps. This focus on the physical world was a clear departure for Rand McNally, though a matter of course in European atlases.

But Goode's desire to communicate complex information conflicted with his more basic concern that American cartography had a tendency to overload the map with information. As he put it, "the map encumbered with useless material . . . is a failure." Students of geography all too often

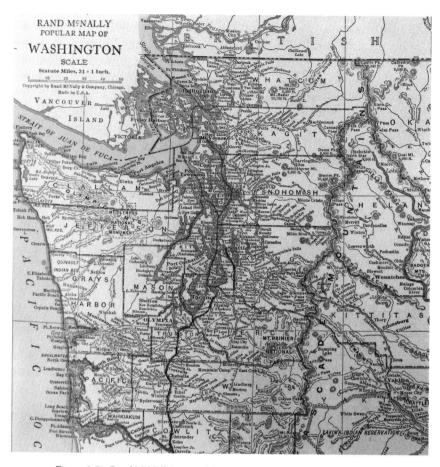

Figure 8.7   Rand McNally's map of the same region as in figure 8.6, marketed to the general public. Courtesy of the Newberry Library.

had to "bend laboriously over the page with a magnifier" to read maps, and in the process lost the relational perspective so essential to the geographical enterprise.[26] To this end Goode designed multiple maps of the same region, each profiling one class of information, such as climate, vegetation, or industrial production (figs. 8.6 and 8.7). The National Geographic Society also began to pull back from this emphasis on quantity of information by touting selectivity of information over comprehensiveness. When introducing its revised map of the world in 1932, the Society defended the decision to print fewer names by explaining that "the ultimate value of the map in many respects depends on the geographer's judgment in choice of place names to be used."[27]

Praise from educators and academics assured *Goode's* atlas an audience within the schools. One reviewer excitedly noted that the maps were "beautifully clear and simple, not obscured by too much detail." And Goode's attempt to regularize scale and proportion of coverage marked a small step toward correcting the "usual confusion about comparative size of continents and countries."[28] Yet many within Rand McNally doubted the extent to which the atlas would interest the public. Andrew McNally was especially skeptical, for though he recognized the scientific superiority of the physical maps, he found himself nostalgic for the more romantic political maps that had become customary representations of the world. As one interviewer wrote,

> a world made up of only slightly varying shades of green and brown hasn't half the appeal (or the romance) of a world of pink and blue and yellow. And while it's nice to know that the city of Manchester is situated (say) ninety feet above sea level, still it is more fun to be able to pick out all over the world the little pink spots of the British Empire.[29]

The political map, dividing the world into empires and nations, had become normative, a kind of metageography. Many of Goode's revolutionary maps and projections were therefore only gradually introduced into the popular atlases. Not until 1937 did Rand McNally include physical relief maps, maps of temperature provinces, and annual rainfall maps in the mass-marketed atlases. Even though it might have presented a more comprehensive picture, *Goode's* world disturbed many at Rand McNally.[30] As one employee commented:

> The total impression, once you get over the shock of a world so grievously sundered, is good . . . political lines are so subordinated to physical features that one is brought up sharply in the realization that . . . France actually does run over into Germany, and Germany into Austria; that one isn't permanently separated from the next by a line and a band of color. But again, the absence of all familiar color makes the book a purely utilitarian object, and not the glamorous gateway to romance that an atlas used to be.[31]

One of Rand McNally's reigning credos had been to create "a harmonious and pleasant looking world." This translated into one divided along political lines, as for many this had become the true representation of the world on a map. The rise of physical mapping challenged the familiarity of the political map that had dominated for decades, and highlighted the degree to which the latter had come to be understood not as a representation of the landscape, but as the landscape itself. In this regard maps are strongest and most persuasive—most scientific and powerful—when they tell consistent messages. Goode's larger project had been to place the United States more fully *in* the world, mapping continuities as well as strict political boundaries. His decision to include fewer place names and more topographical information was part of a larger effort—whether conscious or not—to overturn the tradition created by the gazetteers and atlases of the nineteenth century. Goode was critiquing the characteristics of mass-market cartography that his own publisher—Rand McNally—had been responsible for institutionalizing. More than a decade would pass before Goode's odd-looking maps—though highly successful in American schools —were deemed acceptable for popular consumption and incorporated into the company's general atlases. In the meantime Rand McNally had made clear decisions about the kind of world the public would see.[32]

J. Paul Goode was not alone in his concern with American perceptions, and misperceptions, of geography. *National Geographic* editor Gilbert Grosvenor identified Goode's new homolosine map as clearly superior to the "atrocious" Mercator, but like Andrew McNally found the former visually unappealing. Just after the armistice, Grosvenor asked his chief cartographer to devise a projection that matched the accuracy of Goode's homolosine but maintained the graphic unity of Mercator's world.[33] In doing this Grosvenor had simultaneously recognized both the limits of the Mercator projection and its tremendous visual power. Given his concerns, it is unsurprising that Grosvenor selected the Van der Grinten projection as the basis for the Society's first map of the world in 1922. Designed by Alphons Van der Grinten at the turn of the century, this projection reduced the distortions of the Mercator projection as well as its misleading rectangular shape. But it also retained Mercator's visual appeal by drawing the world on a circle, thereby avoiding the inelegant interruptions characteristic of Goode's design (fig. 8.8).[34] The Society found the Van der Grinten projection worthy of use for nearly all its world maps until 1988, even though it also exaggerated the regions near the Poles, a limitation that, as Jeremy Black has pointed out, was peculiarly appropriate to cold war mapmakers' tendency to emphasize the Soviet Union's power and size.[35]

The Society's decision to use the Van der Grinten projection reflects its dual concerns: correcting for Mercator's well-known flaws while at the same time evoking the texture and shape Mercator had entrenched. Consider the comment made by O. M. Miller, head of the cartographic division of the American Geographical Society, who also tried to devise an alterna-

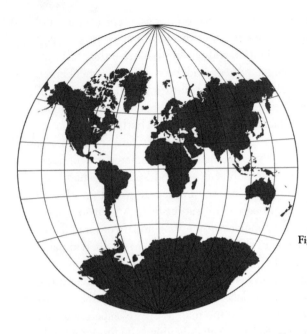

Figure 8.8   The Van der
Grinten projection, used
by the National Geographic
Society as the basis for its
world maps from 1922 to
1988

tive to the Mercator projection. Discussing the parameters of the under-
taking, Miller wrote:

> The practical problem considered here is to find a system of spac-
> ing the parallels of latitude such that an acceptable balance is
> reached between shape and area distortion. By an "acceptable"
> balance is meant one which to the uncritical eye does not obvi-
> ously depart from the familiar shapes of the land areas *as depicted
> by the Mercator projection* but which reduces a real distortion as
> far as possible. . . .[36]

Miller's remark confirms—if unintentionally—the power of the Mercator
projection to be taken *for the world* rather than *as a representation* of the territory. One is reminded of *Tom Sawyer Abroad,* where Tom and Huck drift
eastward across the Midwest in a hot-air balloon. Huck insists they are over
Illinois, for the land appears green, just as it does on the map. Were they
over Indiana, Huck explains, the landscape would surely be pink.[37]

RAND MCNALLY AND THE POSTWAR
CARTOGRAPHIC INDUSTRY

Like the National Geographic Society, Rand McNally initiated aggressive
advertising campaigns to capitalize on the prosperity of the 1920s and the

interest in geography stimulated by the war. But the company's investment in advertising was also a result of increasing competition from other American companies and from Europe. Competition from Europe was a particularly sore subject among American mapmakers, a reminder of the alleged inferiority of their own work. Max Mayer—an American cartographer—sarcastically commented in 1930 that Americans "have had nothing worth the name of an atlas. To claim there is such a publication is to admit our poor aesthetic sense." American atlases, he continued, were simply lost in "a wilderness of meaningless names."[38]

This long-running debate over the aesthetic merits of American maps erupted into a legal and economic question in 1929, when Rand McNally led a campaign to include European maps on the list of imports eligible for duties under the Smoot-Hawley Tariff. To critics like Mayer, the very question of protectionism demonstrated the superiority of European cartography. For Rand McNally, however, the question was more a reflection of its struggle to maintain dominance in America after having adopted new and more expensive physical maps for *Goode's School Atlas*. In fact, the threat came not from European atlases themselves but from the increasingly common practice of American textbook and atlas publishers contracting with European cartographers for maps that were imported and then reproduced in the United States. Though European production methods were more time-consuming, the relative scale of wages in the two countries still favored imports. Companies importing these maps claimed that no comparable product existed in the United States, so faced with increasing losses in their share of the educational atlas market Rand McNally and the Map Engravers and Publishers of America fought to raise the tariff on all European maps. Harry Clow, then president of Rand McNally, charged that old claims about American maps being cheap and inferior products were simply irrelevant and defended the tariff as a way to regulate the disparity in wages between American map drafters, engravers, and printers and their European counterparts. The need for more skilled cartographers to produce *Goode's Atlas* meant higher costs, so Clow framed the tariff as an issue of protecting not a national industry but the creative labor of the American mapmaker.[39]

The most vocal challenge to the tariff came from the Appleton Publishing Company, which introduced an atlas in 1928 that directly competed with Rand McNally's own, especially in its attention to physiography and the problem of projection. Appleton argued that it imported maps from the London Geographical Institute not because they were cheaper but because they were superior to anything available domestically, and that to raise a tariff on these maps would only penalize American schoolchildren in the form of costlier atlases.[40] In response, Andrew McNally II claimed that if

Congress failed to enact these tariffs other publishers would respond as Appleton had, flooding the market with European maps that could potentially put American mapmakers out of business. Rand McNally found a sympathetic ear in Senator Smoot, who repeatedly invoked the perils to the American cartographic industry brought by low tariffs. Others suggested the legislation was simply a way to line the pockets of Rand McNally's executives, and ultimately the tariff remained at 25 percent. With maps that were virtually identical to *Goode's*, Appleton's claim to superiority must have been difficult to accept. Indeed, it appears that Appleton had in fact modeled its atlas on *Goode's*: both included early explanations of scale and projection, and both heavily favored thematic maps of the world's regions. But with nearly twice as many pages of maps in *Goode's*, Rand McNally faced far steeper costs than its competitor.[41]

In hopes of capitalizing on postwar interest in world events—and to protect the market they had previously dominated with relative comfort— Rand McNally also embarked on an advertising campaign that paradoxically celebrated both the romance and the utility of their maps. The association of Rand McNally with accuracy was one of the primary goals of the interwar advertising campaign, and thus the geographic upheavals wrought by the armistice were frequently at the center of these advertisements. One 1921 ad featured a curious young boy asking his father where the newly independent Czechoslovakia could be found on the map. The ad suggested that such a question might easily embarrass any parent who had not kept up with the recent changes, embarrassment that was avoidable through the purchase of a Rand McNally atlas, which could be depended upon to print the latest boundaries wrought by wars or other conflicts.[42] Like the legendary mouthwash ads of the 1920s that preyed on personal anxiety, these advertisements acknowledged the confusing nature of the postwar world and insisted that this knowledge be readily available to every American family. Hammond sold its 1920 *Modern Atlas* through a similar appeal:

See If Your Atlas Shows

The Seat of the League of Nations . . . The Status of the City of
Danzig . . . The New Country of Poland . . . The Plebiscites of
Silesia and Schleswig . . . The New Countries of Czechoslovakia
and Jugoslavia . . . The Empire of Mongolia . . . The Mandatory
Control of Former German Colonies in Africa . . . The Territory
Awarded to France and Belgium . . . The Republic of Esthonia . . .
The Roosevelt River in Brazil.

IF IT DOES NOT SHOW THESE, IT SHOWS

A World That No Longer Exists[43]

Notice here that the focus is still not on geographical relationships but on locations, because for Americans an accurate map was one that identified every city, town, or village. The talisman of comprehensiveness was exemplified by a 1920 ad boasting that Rand McNally maps included villages found nowhere else, including "the little dot that stands for New Dongola." Though few people would ever visit this village, situated between the Sahara and Nubian deserts, they could rely on the company to map its precise location. Whether a reader was looking for New York or Nigeria, Rand McNally promised they would be mapped with equal accuracy, for "maps are worthless unless they are exact." Never was this more clearly demonstrated than in the recent war, "where accuracy meant victory" (fig. 8.9).[44]

These advertising campaigns also suggest that maps and atlases—long marketed as reference tools—were now considered leisure commodities as well, the keys to unlocking the adventures of "Conrad's seas and Kipling's India." Rand McNally atlases would help both children and adults to imagine worlds they might never see; exploring civilizations as old and distant as China, "without stirring from your easy chair." As one advertisement concluded, "every member of your family will profit in culture and knowledge from a Rand McNally Atlas."[45] In previous decades, few if any advertisements or reviews had so directly suggested the cultural value of cartography, stressing instead its utility for the businessman and student. Implicit in this new strategy were evocations of the romantic, adventurous, and even voyeuristic appeal of distant lands. In one ad, the company drew attention to geography through the "Forbidden City of Lhasa," simultaneously compelling and repulsive to Western eyes (fig. 8.10). Rand McNally described Lhasa as a city where

> the Christian is excluded and where decay stalks in the streets. . . .
> The past—with its mystery, its customs, its stand-still civilization,
> lifts its ugly head and leers at the modern and uplifting. Dogs and
> pigs roam at will. The rough lanes are rutted with the traffic of ages.
> Every house is shared by humans and yaks, the common beast of
> burden. . . . This forbidden city has lived for centuries in a little
> world of its own. In the sunlight it is a gorgeous spectacle which
> fades upon close approach into a sordid abode of the unwashed
> and crafty. It is a part of the great romance of Geography, made
> clear by maps. . . . People think of maps when they think of Rand
> McNally—which is a world-known name that stands for educa-
> tion and progress.[46]

This new strategy also reflects the explosive growth of the National Geographic Society in the 1920s. With a membership that reached one million

Where Accuracy Meant Victory

Accurate maps are part of the essential equipment in the enlightenment of the human family.

Accurate maps are the guide-posts of knowledge and the blazed trails of modern business enterprise.

RAND MCNALLY Maps are accurate. They are the products of a highly skilled organization that has made map-making in all its branches a specialty for more than half a century.

RAND MCNALLY Maps are helpful alike to the child in the school, to the student of civic and economic affairs, to the business man developing a market for his goods, to travelers everywhere.

RAND MCNALLY Maps are made in various sizes to meet all needs. Besides their great educational value, they are an indispensable feature of every modern business establishment.

Generals gathered at Staff Headquarters planning a campaign. The fate of men and nations dependent on the accuracy of a map. Here, if ever, a sure test! The result is history.

RAND MCNALLY maps have never rendered more valiant, more vital service than in The World War.

Pershing chose them because of their accuracy, simplicity and quick readability. For the same reasons, they were used by practically every other nation in the war.

There was nothing accidental about this success. It is unusual only in the gravity of the circumstances. Maps and Atlases made by RAND MCNALLY are accurate because we make very definite and sincere efforts to keep them so.

You may never plan a battle or lead a company to the attack. Yet somewhere—*sometime*—you will have need of an accurate map.

Think, then, of RAND MCNALLY and the tremendous scope of its business. Practically every conceivable kind of map for every conceivable purpose is made here at Map Headquarters.

Political maps, Biblical maps, physical maps, climatic maps, historical maps, classical maps, language maps, globes, atlases and Map Systems—these are just a few of the classifications of more than 6000 different *kinds* of RAND MCNALLY maps and atlases.

All of them—each and every one of them—is as accurate and as up-to-the-minute as it is possible for a map to be. RAND MCNALLY has thousands of correspondents in all parts of the world—checking up—seeing to it that not even the least of the world's changes is overlooked.

When you buy a map, buy a good map —a map you can depend on—a RAND MCNALLY!

RAND MCNALLY & COMPANY
Map Headquarters
536 S. CLARK STREET, CHICAGO   42 E. 22ND STREET, NEW YORK

RAND MCNALLY MAPS ARE AMERICAN MADE

Figure 8.9   Rand McNally advertisement from 1920, touting the precision and wartime utility of its maps

by 1926, the Society had clearly tapped the public's desire to learn about the world beyond its borders. With its richly illustrated monthly, the Society brought the distant reaches of exotic lands into American living rooms and libraries. In the process, the Society created a kind of culture around geographic knowledge that could not have gone unnoticed at Rand McNally. In fact, the latter advertised regularly in the pages of the *Geographic*, until in 1928 the Society judged the company a direct competitor and terminated the relationship. These ads suggested precisely the sophistication

The White-washed City
Where Stalks Decay

**The
RAND McNALLY
International Atlas
of the World**

contains 419 pages size
11 x 14 inches closed,
bound in cloth and in full
leather. For the general
reader, student, business
man. 128 pages of maps
covering the entire world
in detail. New countries,
new boundaries, new
groupings of islands.

The reverse side of each
map contains information
about that particular
country or state.

24 pages of full color
illustrations showing in-
teresting and important
sights and scenes in all
parts of the world.

135 pages of indexes
giving location on the
maps and latest popula-
tion figures of practically
all cities, towns and im-
portant places throughout
the world.

RAND McNALLY & CO.

High up, in the mountains of Tibet, is the
forbidden city of Lhasa, where the Christian is
excluded and where decay stalks in the streets,
past the white-washed buildings. The past—
with its mystery, its customs, its stand-still
civilization, lifts its ugly head and leers at the
modern and uplifting. Dogs and pigs roam at
will. The rough lanes are rutted with the traffic
of ages. Every house is shared by humans
and yaks, the common beast of burden.

You'll find Lhasa on the map of Asia, page
225 of the RAND McNALLY INTERNATIONAL ATLAS.
This forbidden city has lived for centuries in a
little world of its own. In the sunlight it is a
gorgeous spectacle which fades upon close
approach into a sordid abode of the unwashed
and crafty. It is a part of the great romance
of Geography, made clear by maps.

No matter what or where the place, you'll
find it on a RAND McNALLY MAP. The uttermost
parts of the earth are included, as they are now
and as they were in olden times. There is a
RAND McNALLY MAP for every person and every
purpose. People think of maps when they think
of RAND McNALLY—which is a world-known
name that stands for education and progress.

RAND McNALLY & COMPANY
Map Headquarters

536 S. CLARK ST., CHICAGO   42 E. 22ND ST., NEW YORK

Figure 8.10 Rand McNally's
appeal to "the romance
of geography" in an
advertisement from 1925.
Courtesy of the Newberry
Library.

and cosmopolitanism that marked those proudly guarding their membership in the Society.[47]

This growth of geography as a leisure activity was also evident in the
industry-wide sales strategies for globes and atlases. *Publishers' Weekly* frequently suggested that bookstores capitalize on current events to sell globes
and atlases. Globes had in fact become fashionable decorative pieces, no
longer simply schoolroom fixtures but now sold in furniture stores and deco-
rating shops.[48] Finally, Rand McNally cultivated product loyalty in the
interwar years through tireless public relations campaigns. By producing
custom maps, globes, and geographic displays for corporate clients and well-
known individuals, the company kept its name in the public eye and solidified its cartographic reputation. American Airways, Texaco Oil, CBS

Figure 8.11   Public map posted by Rand McNally, New York. Courtesy of the Newberry Library.

Radio, the Bureau of Reclamation, International Harvester, National Cash Register, and the *Christian Science Monitor* were just a few of the organizations that contracted with Rand McNally to build public maps and globes for company lobbies, expositions, store windows, and railway terminals. Dozens of these projects helped build the company's reputation as a ubiquitous, reliable, and authoritative source of knowledge about the newly configured world between the wars (fig. 8.11).[49]

The early twentieth century brought a slow and steady stream of Americans into contact with maps and atlases. Geography was gradually becoming not just a school subject or a reference tool, but a cultural commodity as well. Rand McNally translated this interest into a broadened audience for cartography, aided particularly by the booming demand in the 1920s for domestic road maps. This ability to strengthen its reputation as a cartographic authority would prove central to the company's success in the 1940s. Like the National Geographic Society, Rand McNally was successful because it strove to create a world that made sense to its public. Though it capitalized on the upheavals brought by World War I and the Treaty of Versailles, the company also chose to present a world that fit American notions of how the world ought to appear. In fact, what is most striking about these atlases is the degree to which they maintained a tradition begun years earlier. Rand McNally kept new maps out of mass-market atlases in the 1920s because the company was wary of directly challenging a public whose visual sense of the world had been cultivated by years of exposure to certain types of maps. However appropriate the unfamiliar maps were for "educational" purposes, Rand McNally was cautious about their acceptability as products for mass consumption. This suggests that the history of cartography is governed not just by technological and scientific advances, but also by a complex interplay of expectations between mapmakers and consumers. While the atlas remained relatively stable through the early interwar years, by the 1930s news of conflict in Europe and East Asia once again drew American eyes abroad, and the war that followed, together with the revolution in aviation, challenged both geographic realities and their cartographic representations.

9

War and the Re-creation of the World
1939–1950

On Friday, February 20, 1942, President Roosevelt asked Americans to buy a map of the world. In his noontime radio address Roosevelt announced that he would explain the nation's wartime strategy over the airwaves the following Monday, and that a clear sense of geography would greatly facilitate this task. That weekend map sellers experienced a rush of business, oil companies announced map giveaways at service stations, department stores advertised their maps in the Saturday papers, and nearly every newspaper published a world map in their Monday edition. Some movie theaters even planned to show a world map onscreen while broadcasting the President's message. That Monday, in his first fireside chat in more than two months, millions of listeners complied when Roosevelt asked them to "take out and spread before you a map of the whole earth."[1]

Over the course of the war, this nationwide attention to cartography brought the farthest reaches of the world into everyday conversation. Yet the fact that the war raised public awareness of geography is fairly unremarkable: war has always generated popular interest in geography. Far more significant is that Americans were buying and using maps that were unlike anything they had ever seen, and that these shocking new images came to them not through the established channels such as Rand McNally, but from other sources. While commercial map companies continued to sell standard renditions of the world, newer maps showed the public that geography had been utterly transformed by the exigencies of war and the technological advances of aviation. As a result, during the war Americans for the first time confronted competing visions of the world, brought into their homes on an unprecedented scale and in an unprecedented manner. This proliferation of new and different maps sparked a vigorous public dialogue—carried on in magazines, newspapers, radio shows, and newsreels—about the nature of cartography, and of geography generally. Ironically, this heightened attention illustrated just how entrenched American views of the world had become, and how long the established map companies had enjoyed unchallenged reputations of authority and accuracy.

One of the few historians to explore the meaning of maps in American culture is Alan Henrikson. In the 1940s, Henrikson argues, the anticipated age of aviation, together with the circumstances of a truly global war, created a new view of international relationships, one he terms "air age globalism." The bombing of Pearl Harbor officially brought Americans into the war, but it also abruptly initiated a new era of geography. The fact that Americans were now fighting an aerial war—to the east, to the west, and potentially to the north—effectively eroded any sense of hemispheric isolation, and encouraged instead a sense of vulnerability both over land and across oceans. As Henrikson writes, aviation demonstrated that "distance has human significance only as it affects communication, travel, and transport." Stated by a wartime observer in 1944, "space alone has no significance . . . it is mobility in space that gives it meaning."[2] This geographical reorientation demanded a fresh recognition of the earth's spherical shape, seriously challenged the mental separation between Western and Eastern hemispheres, and drew the attention of the public to the Arctic as an arena of strategy. Significantly, none of these new geographical "realities" generated by the war could be illustrated by the maps commonly found in circulation. This was especially true for the Mercator projection, which translates the globe into a cylinder. Though this map preserves directional accuracy—an advantage particularly important for navigation —it exaggerates substantially the northern and southern latitudes (see fig. 8.4). The rapid growth of aviation technology in the 1930s and 1940s made this projection's limits clear in three ways. First, oceans, whether the vast seas to the east and west or the frozen Arctic to the north, no longer held great power as physical barriers. Second, it was impossible to chart air routes accurately on the Mercator projection, because the great circle routes could not be easily drawn. Finally, the traditional projection could not illustrate the centrality of the North Pole; on most versions, this region did not appear at all.

Though this reorientation of perspective came to the majority of Americans only during the war, the influence of aviation over ideas of geography had in fact been demonstrated a few years earlier. In 1937 two Russian pilots flew from Moscow to America over the North Pole, an achievement that won the hearts of Russians and the admiration of many around the world. Vilhjalmur Stefansson, the legendary arctic explorer, quipped that a monument to these pilots ought to read "they found the world of transportation a cylinder; they left it a sphere."[3] As John Mc-Cannon writes, this flight was a mastery over the "highest of high grounds," and the Russian polar feats of the 1930s "involved the uncovering of the earth's final, most jealously guarded secret, the last 'white spot' on the map."[4] Yet, significantly, these achievements did not immediately transform public depictions of space. Instead, it was the magnitude of war a few years later that forced this reconfiguration. During the 1940s, as maps emphasized—and occasionally exaggerated—the possibilities of air travel, American policy makers were sensitized to the nation's vulnerability relative to the North Pole, which Henrikson considers a contributing factor to the cold war.

Henrikson's exploration of the influence of maps on foreign policy—now more than twenty years old—raises other questions of how maps were produced and disseminated in American culture, and what messages they sent to a public fighting a global battle. Prior to the war, American cartography was a remarkably static industry that for decades had marketed to the public a single vision of the world. The wartime craze for maps and geography forced the nation's largest map producers to recreate their maps and world atlases, a belated response to a consumer market that had been exposed to a qualitatively more sophisticated and flexible sense of geography. Maps were redesigned not just to reflect changing boundaries wrought by the war, but also because the public had paid sustained attention to innovative maps produced by nontraditional mapmakers, thereby turning the map itself into contested ground. Together these changes make the 1940s a significant moment in the culture of geography.

"War Is God's Way of Teaching Us Geography"

War has perennially boosted popular interest in geography. The Spanish-American War and World War I boosted map and atlas sales and focused attention on the place of geography in American education. Maps and atlases made by the largest companies—Rand McNally, C. S. Hammond & Company, and the G. F. Cram Company—had grown as a leisure market throughout the 1920s and 1930s, fueled by the construction of new roads, the auto industry, and the rise of an American consumer culture. In the Second World War Americans again sought out the map, but this time on a much larger scale. After the Nazi invasion of Poland, demand spiraled until maps and cheap atlases were ubiquitous, available in newspapers and magazines, bookstores and department stores, press rooms and living rooms, train stations and gas stations.

On the first of September, 1939, the Nazis invaded Poland, and by the end of the day a map of Europe could not be bought anywhere in the United States. That week Rand McNally reported selling its entire stock of European maps and a sizable share of its atlases. New Yorkers grabbed atlases off the shelves of Brentano's, Scribner's, and Doubleday, while broadcasting stations, magazines, and newspaper offices scrambled to find any decent map of the area. Responding to the demand, book and map retailers across the country immediately devoted their window displays to the invasion (fig. 9.1). Rand McNally reported selling more maps and atlases of the European war regions in the first two weeks of September than in all the years since the armistice of 1918. In that same period the Hammond Company sold 300,000 maps of Europe, an astronomical figure compared to their weekly average sales of a few dozen. Sales continued to be strong throughout the fall of 1939 and all of 1940, boosted by the Nazi invasion of Belgium in May.[5] Predictably, the attack on Pearl Harbor again brought an enormous demand for maps. A supply of fifty-cent Rand McNally Oceania maps expected to last the entire year sold out in a single day. On December 10 the company printed 15,000 copies of a world map and 15,000 of a Pacific map, all of which sold by the following Friday. By Christmas, the supply of atlases, globes, and maps of the world and the Pacific was virtually exhausted, and both Hammond and Rand McNally reported their largest annual sales to date in 1941.[6]

Because of the scale of the conflict, and because of American commitments, these atlases and maps were used in a more participatory manner than those released during World War I. In an atlas given away through the mail and cosponsored by Planters Peanuts, Rand McNally urged readers to "Watch Those Axis Losses Grow! Keep your own record as retribution comes to Germany and Japan," and thus included columns to track the loss of enemy men and materiel. One advertisement urged consumers to "Follow the Yanks' Victory March" by tracing the European and Asian war campaigns with markers to identify troop movements, air raids, naval battles, and the like.[7] World War I had demonstrated that the public enjoyed following the campaigns on highly localized maps, so Rand McNally and Hammond frequently included pins, small flags, and thumbtacks with which to mark the shifting tides of battle. As one writer commented, "it's a grim sort of game, but a lot of people apparently like to play it."[8]

Despite the massive distribution and creative packaging efforts, however, these maps and atlases were decidedly unsuited to following the war. Though billed as "war" atlases, few included maps that had been redesigned

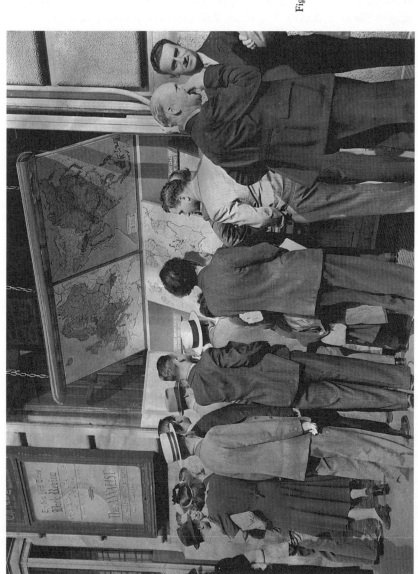

Figure 9.1 The news of war in Europe grips Americans and sparks a run on maps. Here the public refers to the world map in the window of the New York Times Company, September 9, 1939. NYT Pictures.

for the circumstances of armed conflict. Rarely did they mark the latest boundaries or military relationships; instead they continued to include the detailed, densely lettered locational information that had been character-istic of American cartography for decades. Those atlases specifically de-signed to follow the war campaigns included little about the actual contours of the fighting, or even the position of allies. By showing so many place names, rather than selectively illustrating relationships, the maps failed to convey the nature or magnitude of change brought by war.[9] In fact, Rand McNally and Hammond's peacetime atlases had changed little over the first half of the twentieth century, a function of the technological con-straints of map printing but also of the stylistic inflexibility of American map companies. Thus, while the war gave established mapmakers a captive audience, their maps only partially fulfilled the need. The war necessitated maps of singular and immediate relationships between locations and troops, and this meant departing from conventions of style, projection, and per-spective. Thus the limited utility of the traditional map became undeni-able during the war.

The National Geographic Society printed more than 37 million re-gional maps during the war. Little wonder then that so many Americans recall following the news on *National Geographic* maps tacked to the walls of their living room, kitchen, or library.[10] The Society had published maps occasionally since the turn of the century, but in the 1930s their produc-tion became routine, and from 1936 to 1947 thirty-eight maps were is-sued. The extraordinary popularity of these maps had much to do with the Society's unique position in American culture. Distributed as supplements to the *National Geographic Magazine*—which had a circulation by 1940 of 1.1 million—the maps were guaranteed a broad audience. But this does not explain the demand for these maps beyond their subscribers, which has to do with the Society's reputation in American culture and also its carto-graphic innovations.

For instance, only two of the Society's thirty-eight wartime maps used the traditional Mercator projection. This willingness to experiment with more than a dozen other projections directly challenged assumptions of how the world ought to look. Furthermore, because they were issued indi-vidually rather than in an atlas, these maps could be tailored to fit the loca-tions and situations dictated by the day's headlines. More attention could be paid to the particular circumstances of each region, which in turn al-lowed for the inclusion of more relevant detail on the map. For instance, the Society's map of Japan—issued in April 1944—was drawn on a special projection centered on Tokyo. This deliberate choice allowed the reader to trace bomber courses as straight lines outward, true in both distance and direction, a quality that made the map popular among civilians as well as military personnel. With its well-respected name, the Society was able to experiment with unfamiliar cartographic ideas and views while maintaining its reputation for accuracy and reliability.[11]

The Society imaginatively identified on the map not just place names that were newsworthy, but also those that captured the mind of the adventurous reader: one map of the Atlantic Ocean issued in September 1941 identified the places and dates of German submarine seizures as well as seventeenth-century voyages. Unlike the Rand McNally, Cram, and Hammond maps, which were generally designed as reference sources, these maps were interesting artifacts in and of themselves that contained stories as well as information. As discussed in chapter 8, the Society had begun to reduce the number of place names in the 1930s, when its cartographers realized that selectivity allowed for more intriguing and compelling detail. As a cartographic consultant for the rival Hammond Company himself recognized, the Society's maps had begun to achieve a balance of detail and elegance that eluded commercial mapmakers. This clarity was also due in part to the Society's advancing print techniques. The Society's chief cartographer, Albert Holt Bumstead, developed a method where each letter of the alphabet was drawn and then photographed. In this way the lettering could be reproduced in any size without sacrificing legibility; by the late 1930s all of the Society's maps were produced using this technique (figs. 9.2 and 9.3). The Society's cartographers also used color differently, forgoing the traditional heavy layer tints that had been used to mark regions and countries in favor of neutral colors for backgrounds and darker shades to demarcate boundaries, rivers, highways, and the like. This greatly facilitated the identification of place names.[12] These changes, along with other improvements, went a long way toward improving the aesthetic quality of American-made maps, which until then had been considered markedly inferior to the European variety.

All this care in making the map was not lost on the Society's patrons, whose testimonials—especially those from the military—were dutifully recorded. General Eisenhower was particularly impressed with the Society's July 1944 map of Germany, which was used to make 20,000 enlargements for the Ground and Air Forces in Europe. Early in 1942 *Newsweek* dubbed Washington "a city of maps" where "it is now considered a *faux pas* to be caught without your Pacific arena," and it was the Society's Pacific map that was in high demand at the State Department, the White House, embassies, and among the military.[13] The Air Force used the original drawings of the wartime map of Asia to make enlargements for its air offensive against the

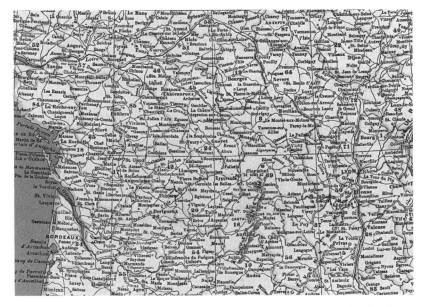

Figure 9.2   Detail of Rand McNally's wax-engraved map of France produced in the 1940s. Compare with figure 9.3. Courtesy of the Harvard Map Collection.

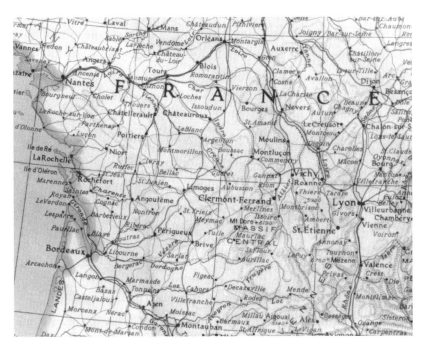

Figure 9.3   Detail of *National Geographic*'s map of France, also produced in the 1940s. Notice the clarity in appearance brought by the newer technique of photocomposition, aided greatly by the decision to include fewer place names.

Japanese. Eighty thousand copies of the China map originally published in the June 1945 *Geographic* went to fill one War Department order alone.[14] After the war, Chester Nimitz, Commander in Chief of the Pacific Fleet, recounted the pivotal role a Society map had played in one of his missions. Finding himself and his crew lost over the South Pacific, one officer's habit of carrying National Geographic maps helped the company safely navigate its way to Guadalcanal.[15] The Society's maps also were used to prepare soldiers both stateside and in the battlefield for the geography of the war. Approximately 250,000 of these *Newsmaps* were distributed to active Army and Navy servicemen around the world for instructional purposes. Even more highly circulated were the National Geographic maps reprinted in *Yank*, an Army weekly for soldiers with a circulation of three million. All told, the War Department ordered more than one million maps from the National Geographic Society for use at all levels of combat and strategy (figs. 9.4 and 9.5).[16]

Figure 9.4   French girls following General Patton's maneuvers on a National
Geographic Society map [courtesy National Archives/U.S. Army Signal Corps
SC2018175]

Figure 9.5   American military officials consulting the National Geographic Map of Africa [courtesy National Archives/U.S Army Signal Corps SC163930]

The Society's maps also became fixtures in the media as reference sources for reporters and as models for journalistic cartographers. As the *New York Times* reported in 1942:

> Inevitably, the newspapers have had to lean on many sources to make intelligible the brief and sometimes cryptic dispatches and communiques. And chief among them is the old reliable National Geographic Society. Not only has it given generously of its information and opened its magnificent geographic library; it has published splendid war maps and it issues daily bulletins to assist the newspapers and news services.
>
> It is highly reassuring, when such place names as Staryi Oskol and Zivotin crop up in the communiques, to be able to ask somebody what and where they are—and get the right answers. The N.G.S. hasn't failed us yet.[17]

The influence of the National Geographic Society's maps was exceeded only by those printed in the nation's newspapers and magazines. Journalistic cartography had a long tradition in the United States, debuting in the

Civil War and reappearing in scattered form in World War I. But only in World War II, when exposure to newspaper and magazine maps reached massive, daily proportions, did they become a force to be reckoned with. The number of maps in the *New York Times*, for instance, rose precipitously during the 1930s and 1940s as a result of the war and the introduction of a daily weather map in 1934. Advances in print technology also facilitated this increase, and by 1950 the number of news maps related to defense and geopolitical themes reached a level unmatched either before or since.[18] Journalistic cartographers were bringing maps to the public in greater numbers than ever before, but just as important was the experimental style of these maps. A host of amateur cartographers at newspapers such as *The Christian Science Monitor*, *The New York Times*, *The Chicago Tribune*, *The New York Herald-Tribune*, *The Milwaukee Journal*, and *The New York Daily News* experimented with perspective, color, and projection to illustrate the war for their readers. These maps also flooded weeklies and monthlies such as *Time*, *Fortune*, and *Life*, as well as the Luce publications, *Newsweek* and *Collier's*. One contemporary cartographer, Walter Ristow, praised these journalistic cartographers for bringing creativity and artistry back into the industry, even though many of them had little or no formal training in mapmaking. With their brash aesthetic style these maps demonstrated how static the commercial maps had become by comparison.[19]

RICHARD EDES HARRISON AND THE CHALLENGE TO AMERICAN CARTOGRAPHY

The simple, dramatic style of newspaper cartographers owed much to the work of Richard Edes Harrison, the individual most responsible for sensitizing the public to geography during the war. His hand-drawn maps — appearing primarily in *Fortune* magazine and endlessly reprinted and copied elsewhere — simultaneously shocked and dazzled readers, showing them the world in terms of relationships that were left hidden on more traditional maps.[20] A master of self-promotion, Harrison distinguished himself from established cartographers by touting his training as an architect and identifying himself as an artist rather than a mapmaker. Harrison excoriated professional cartographers for their failure to explain to the public the spatial realities of the war, a task that was "assumed by rank outsiders — the magazines and daily papers."[21] It was his own lack of formal cartographic training, Harrison argued, that enabled him to experiment with the narration of space.

To look at Harrison's maps is to see the landscape in a very specific way. Focusing on the revolution brought by the war and aviation, his maps resemble a photograph of the earth from a distance. Through this perspective Harrison translated a three-dimensional view of the air-age world onto a two-dimensional plane, creating a sense of globularity that ordinary commercial maps could not match. For a culture reared on the rigidity of the Mercator projection the visual impact of Harrison's maps should not be underestimated. The powerful simplicity of this perspective silently—yet insistently—forced the reader to conclude that the world had been reshaped through the advent of aviation. Harrison's use of perspective, together with his decision to include only those place names that were necessary to understand the argument of the map, tailored his perspective maps to the singular purposes of the crisis.

The visual power of Harrison's maps was brought home vividly in his "Three Approaches to the United States," one of a series of maps published in the *Fortune* issue of September 1940 as an "Atlas for the U.S. Citizen" (see fig. 9.6). In this map Harrison brought the distant war home to America not across the Atlantic to New York but to Detroit (from Berlin), to Seattle, Salt Lake City and Denver (from Tokyo), and to the "soft belly" of the American south (from South America). As Harrison recalled, the maps were designed to prove to Americans that "however remote Europe and China seemed, we were part of that world and not as remotely as many of our good citizens thought."[22] But while the war in Europe was already a year old, Americans still maintained a strong sense of graphic isolationism, reflected by the fact that three different publishers rejected the proposal to publish the maps together as a separate atlas for the general public. After the bombing of Pearl Harbor, however, Harrison's importance as a prophetic cartographer skyrocketed, and the editors at *Fortune* continued to print his maps throughout the war. Many, such as his "One World—One War" supplement of March 1942, became wall maps in American homes and schools (fig. 9.7). This map made tangible an argument that had been quietly circulating for years: that aviation had transformed the North Pole from a barren wasteland into an arena for communication and conflict. First printed in the August 1941 *Fortune*, in an issue devoted to the possibility of American entrance into the war, this polar map was intended by Harrison to demonstrate that North America was central, rather than peripheral, to the conflict. The text accompanying the map underscored the argument that, in terms of ideology, geography, and the new policy of Lend Lease, the "entire conflict pivots around the U.S." Although the map separated the nations of the world into seven camps, there were only two that mattered: "those who are for us . . . and those who are against us."[23] Like the National Geographic map of Japan, this was drawn on a projection that allowed one to trace directional relationships over the Poles. As with any

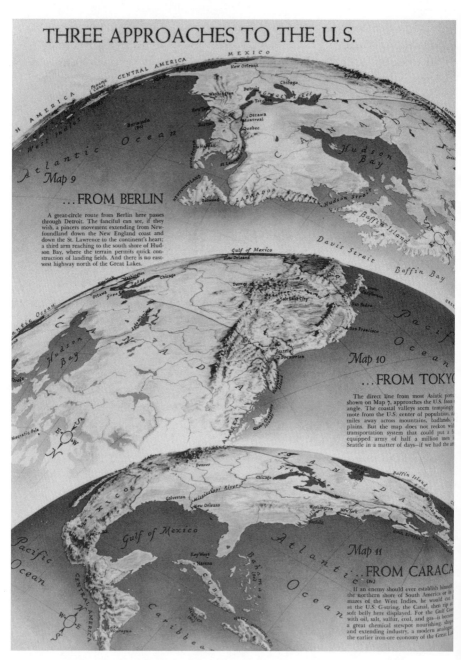

THREE APPROACHES TO THE U. S.

Map 9

...FROM BERLIN

A great-circle route from Berlin here passes through Detroit. The fanciful can see, if they wish, a pincers movement extending from Newfoundland down the New England coast and down the St. Lawrence to the continent's heart; a third arm reaching to the south shore of Hudson Bay, where the terrain permits quick construction of landing fields. And there is no east-west highway north of the Great Lakes.

Map 10

...FROM TOKYO

The direct line from most Asiatic ports shown on Map 7, approaches the U.S. from angle. The coastal valleys seem tempting mote from the U.S. center of population, miles away across mountains, badlands, plains. But the map does not reckon wit transportation system that could put a equipped army of half a million men Seattle in a matter of days—if we had the an

Map 11

...FROM CARACA

If an enemy should ever establish himself the northern shore of South America or in mazes of the West Indies, he would com at the U.S. G-string, the Canal, then rip soft belly here displayed. For the Gulf Co with oil, salt, sulfur, coal, and gas—is bec a great chemical stewpot nourishing, shap and extending industry, a modern analog the earlier iron-ore economy of the Great La

Figure 9.6 "Three Approaches to the United States," designed by Harrison and published in *Fortune*. Harrison intended these three maps to demonstrate the vulnerability of the United States from all directions.

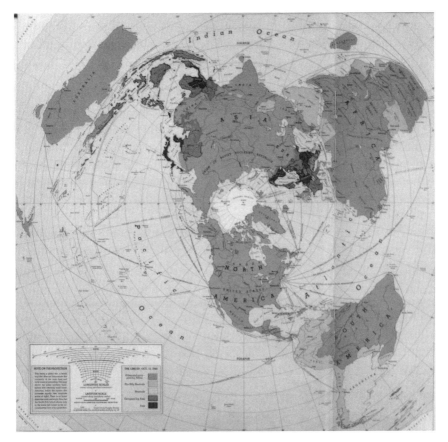

Figure 9.7   Harrison's "One World, One War" map drawn on the polar azimuthal
projection

world map, accuracy in one area came at the expense of inaccuracy elsewhere, and here the southern continents were distorted beyond recognition. Yet this was a minor disadvantage during a war staged in the northern hemisphere, and the limits of the polar map seemed to the public a
small price to pay for the larger truths it revealed. As argued in chapter 6,
American schools were particularly receptive to this new cartographic
model for its ability to portray the world as a group of tightly wedged continents brought closer by the dramatic advances in aviation.

When a collection of his maps were published as an atlas in the spring
of 1944, Harrison's influence extended even further. The maps in *Look at the
World* forced a reevaluation of assumptions about the appearance and shape
of the world, about distance and direction, and about America's role abroad.

Harrison geared *Look* toward a visualization of the relationship between cities and regions, such as the Middle East from Europe, or western Europe from Germany. Especially startling was his view of Japan from Alaska, which brought Japan virtually to America's western doorstep, a relationship illustrated poorly—if at all—by the Mercator projection (fig. 9.8). The use of a polar route to connect Japan to Alaska effectively transformed the Pacific from a massive body of water protecting the United States into a smallish lake. The shrunken ocean connected the American nations to Asia rather than separating them, linking the Eastern with the Western hemisphere. In a similar fashion, Harrison's map of "Japan from the Solomons" highlighted the proximity of the two island chains in a way that remains hidden on traditional interwar maps of Oceania, generally drawn on the Mercator projection (figs. 9.9 and 9.10).[24]

Most Americans who wrote to Harrison conveyed a deeply internationalist sensibility, though whether his maps generated or reinforced these ideals is unclear. Harrison's writings in the 1940s, often coauthored with foreign policy experts, urged Americans to take up the challenge of world leadership. But this broadly internationalist message carried an array of possible readings. Harrison drew some Americans closer to the conflict and closer to the enemy, heightening a sense not only of national power but also of national vulnerability. Alexander De Seversky's *Victory through Air Power*, a popular Book-of-the-Month Club selection in 1943, captured perfectly the fears brought by this new sense of geography.

> From every point of the compass—across the two oceans and
> across the two Poles—giant bombers, each protected by its con-
> voy of deadly fighter planes, converge upon the United States of
> America. . . . Aerial armadas now battle boldly and fiercely, just as
> great naval armadas used to do in the past, only with a destructive
> fury infinitely more terrifying.[25]

To be sure, De Seversky was exaggerating; the book even included a disclaimer from the Book-of-the-Month Club so that readers would not be alarmed by his conclusions. But his aim was to impress upon Americans the magnitude of the geographical change brought by war and aviation. Here and in his subsequent writings De Seversky used orthographic and pole-centered maps to demonstrate the need for the United States to defend itself against foreign threats from the skies and over the Pole.

But Harrison's maps could also motivate a strain of internationalism that was highly humanistic. During and after the war, writers such as Wallace Stegner, Wendell Willkie, Edward Steichen, and Carey McWilliams were introducing a new transnational identity to Americans through their

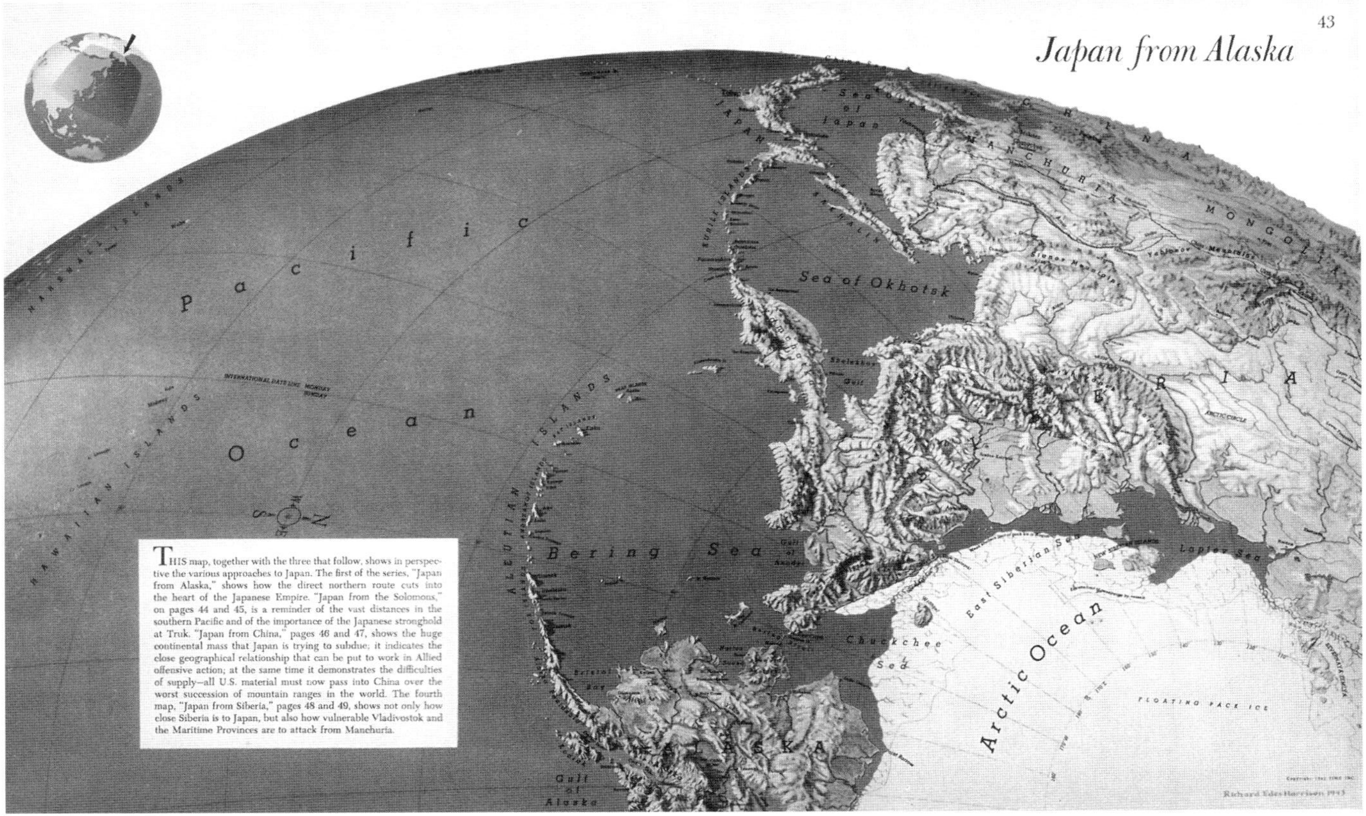

Figure 9.8 Harrison perspective map, depicting the relationship between Japan and Alaska

Figure 9.9 Rand McNally's traditional interwar view of the Pacific. Compare with perspective in figure 9.10. Courtesy of the Harvard Map Collection.

Japan from the Solomons

Figure 9.10   Harrison's view of Japan from the Solomon Islands

textbooks, photography, and other media. This new identity, I would argue, was closely related to Harrison's perspectives. Consider that Willkie's *One World*, written in the midst of the war, opened with an air-age map charting his forty-nine-day flight around the world. The map, together with his narrative, implied a world of easy internationalism. Willkie spoke of men and women from different parts of the world as if they were his neighbors. He crossed racial, ethnic, and national lines to impress on Americans the *dangers* of nationalism in a world so tightly woven together. Willkie's description of this new world included a number of references to the spatial reorientation brought by war, aviation, and technology. It should be unsurprising that Willkie also used a Harrison-style map to represent his optimism about the future. Similarly, just after the war the *New York Times* commissioned Harrison to draw one of his trademark perspective maps to cover the wall of its midtown Manhattan lobby, a massive image that remained in place for twenty-five years. Most striking about the map is the accompanying quotation, taken from a late nineteenth-century poem entitled "Begin Again": "Every day is a fresh beginning / Every morn is the world made new." The choice of so optimistic a verse to describe the map reflects how Harrison's work could be read in a highly humanistic manner, a dramatic contrast with the tone of De Seversky's geopolitical writings. The maps could reinforce national divisions by showing the relationship between belligerents, but simultaneously could suggest that national borders created a *false* sense of isolation. In both cases, the overriding message was a sense of immediacy that was completely obscured by the Mercator projection.

Harrison's maps—like those of the National Geographic Society—were widely distributed within the military for both educational and strategic purposes. With civilians, Harrison made an even stronger debut. The first edition of his *Look at the World* atlas—long anticipated by the media—sold out before even reaching the bookstores, and Harrison was deluged with map requests from companies, citizens, and civic organizations hoping to put his perspective to special use. Lockheed Aircraft was so impressed with Harrison's polar projection that it requested a similar map with each of the continents placed at the center.[26] The editors of the *Atlantic City Press/Evening Union* asked for an azimuthal projection centered on Atlantic City to show readers the relationship of New Jersey to the world beyond. Maxwell Benson, on behalf of the Nashville Community and War Chest, wrote to ask Harrison's advice on choosing a map to educate Tennesseans about their intimate connection to the nations receiving aid from the war chest. As Benson explained to Harrison, he saw "the greatest need for visualizing to the public where this money goes; whom it helps. . . . Local people have got to be made to see this need globularly [*sic*]." Benson's letter, like dozens of others sent to Harrison, reflect the latter's ability to shape — and reshape — the geographical sensibilities of the general public.

Most professional geographers and cartographers who wrote to Harrison were ardent supporters of his work. Erwin Raisz, a cartographer who enjoyed tremendous acclaim in his own right, enthusiastically encouraged Harrison's cartographic style and drew upon it freely to educate his own students at Harvard. Others were less sanguine. Wellman Chamberlin, staff cartographer at the National Geographic Society, judged Harrison's maps artistic rather than cartographic for sacrificing mathematical precision and conformity of shape in order to convey a three-dimensional relationship. Chamberlin classified Harrison's work as propagandistic, and compared it to the function and style of cartography in Nazi Germany. Both, he argued, subordinated accuracy to dramatic illustration. Chamberlin saw a wide gulf between "propaganda" maps and what he considered the "objective" mapping pursued by the National Geographic Society and other legitimate cartographic authorities.[27] Another vigorous critic was Charles Colby, chairman of the geography department at the University of Chicago, who refused even to accept Harrison's work, "messy in appearance and confused in detail," as cartographic. Harrison's failure to conform to conventions — such as the precise use of coordinates and projections or the accepted orientation of a north-south axis — meant that his maps "were not maps at all."[28] In response, Harrison reminded critics like Chamberlin and Colby that one of the paramount goals of a map was to elucidate geographical relationships, and that all maps sacrificed some precision in their attempt to represent the earth graphically. Harrison even insisted, well before such a claim became commonplace, that all cartography was inherently argumentative, and all maps inherently suggestive. As he wrote to a colleague,

> the representative of one of the big map companies came to me and sounded me out on the possibility of my drawing a couple of hemispheres for a sort of wall atlas which they are about to produce. What he revealed to me — unwittingly perhaps — was the extraordinary timidity of map companies in regard to trying anything which has the slightest tinge of unorthodoxy. He even said they couldn't present an orthographic projection of the Eastern Hemisphere centered anywhere but on the Equator. He gave me several examples of how staid educational directors and professors

with influence in the teaching of geography are able to dictate to the map companies and keep them in the strict groove of carto-graphical orthodoxy, including of course the greatest sacred cow of all: north should always be at the top of the page. Judging from the reader reaction to Fortune maps, some of which are admit-tedly whacky, and also the increasingly flexible use of maps by advertisers, the general public is way ahead of teachers and map companies.[29]

Harrison's iconoclastic maps paved the way for the work of dozens of other journalistic cartographers across the country who exposed untold numbers to unusual maps on a daily basis throughout the war. The newspaper and magazine maps of *Time*, *Newsweek*, and the Associated Press were also gathered into atlases, offering consumers an array of options that flouted cartographic conventions. The editors of the *Time* atlas used bold maps to identify economic resources and drew boundaries along occupational rather than sovereign lines to illustrate areas crucial to military strategy and their relationship to political borders. The maps of the Balkans and the Ukraine graphically illustrated the physical contours of the land, a method popu-larized by Harrison as shown in figure 9.11, bringing home the strategy of the war and actually pulling the reader into the landscape, thereby inten-sifying his engagement with the map and the battle. The description of one of *Time*'s maps strengthened its message: the East Indies was an area "where Japan's new order is nudging closer and closer to the Philippines—where a dozen little countries and orphaned colonies cower before the advancing shadow of totalitarianism . . . where the U.S. fleet may soon be cruising the waters of the spice islands and the Rising Sun." The *Newsweek* war atlas took this one step further by using Harrison-style maps to envision the German and Japanese perspectives in the war, showing both their attacks outward and attacks upon them. These compelling maps broke the carto-graphic logjam and widened the popular imagination during the war.[30]

The market share enjoyed by Rand McNally and Hammond in the 1930s meant that their products heavily influenced the American public's idea of what maps and atlases *ought* to look like. But while the war and the air age ensured a market for maps, this was so only insofar as these prod-ucts addressed contemporary issues and debates. Americans were drawn to other maps in part because they rejected the technical style of cartography in favor of a clearer, more visceral approach that suited the crisis. Though some accused Harrison of work that was more propagandistic than carto-graphic, the very availability of diverse maps suggested that *all* maps are arguments and that none can be understood as unmediated. The moral

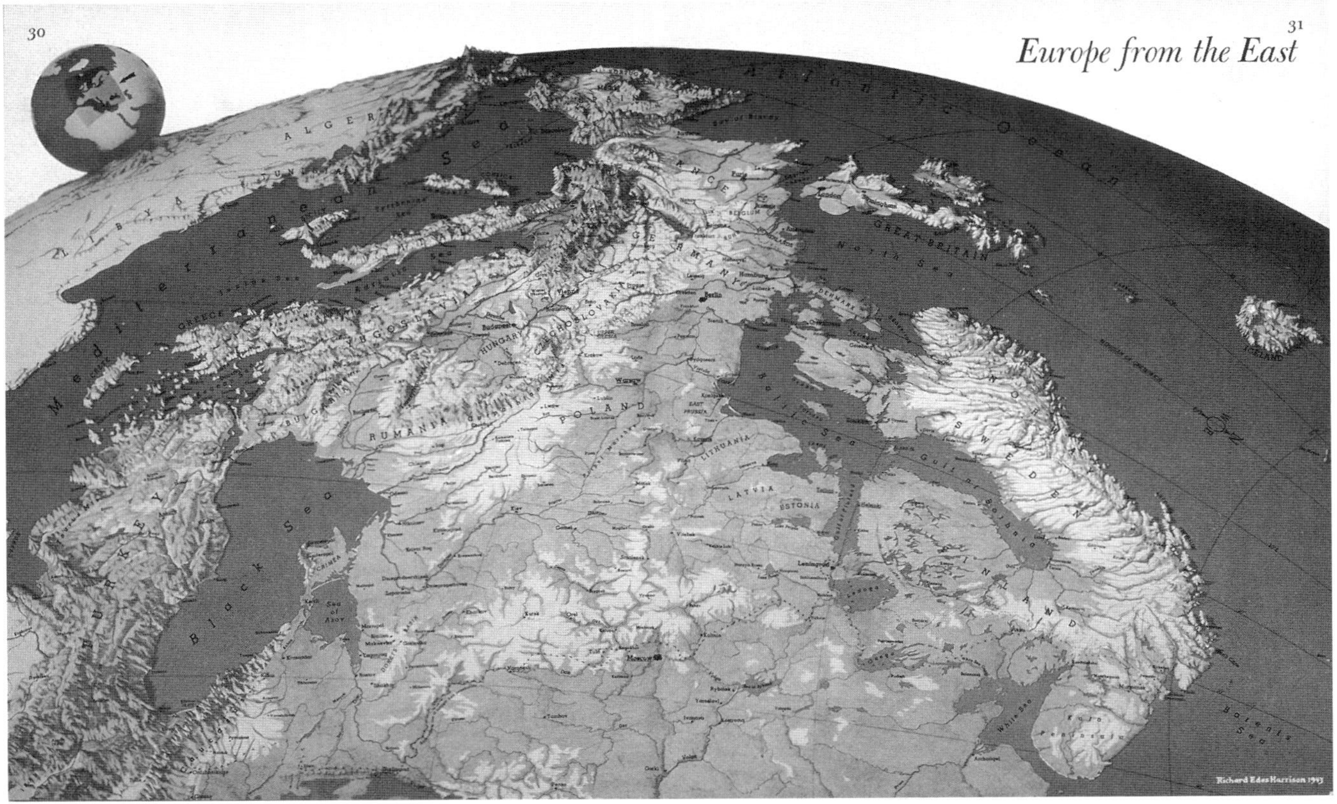

Figure 9.11 Harrison's view of "Europe from the East," which translates the two-dimensional traditional map of the region into a three-dimensional "bird's eye" view of the battle regions

authority of the Allied cause actually enabled journalistic cartographers to embrace an argumentative posture; in terms of style these maps even began to resemble advertisements or war posters.[31]

A Reorientation of Space

During the 1940s geography became part of the national culture, surfacing in magazines, newspapers, and radio shows, and taking center stage in the schools. As one Rand McNally cartographer excitedly—perhaps hyperbolically—commented, before the war the "average man's" knowledge of places was woefully limited:

> Hawaii was associated with surfboards and the beach at Waikiki; Bali was a beautiful island inhabited by still more beautiful dreams; the South Sea Islands were bits of paradise where food grew upon the trees and no one ever worked; New Guinea was a place where missionaries were sent, and comparatively few individuals had the remotest idea of the location of any of them. . . . Today, however, with a global war in progress and American fighting men stationed in more than fifty countries or colonies, we find an entirely different story. All about us people are tossing about such names as Guadalcanal, Attu, Pantelleria, and other tiny places with the greatest of ease and familiarity, while in restaurants we find armchair strategists capable of sketching very creditable maps of the Soviet Union on napkins and even spelling such names as Dnepropetrovsk and Simferopol correctly.[32]

Yet the maps discussed above indicate that the public's fascination with geography was compounded by the sense that space itself was being fundamentally altered. During his fireside chat of February 1942, for example, Roosevelt described the Atlantic and the Pacific not as massive buffer zones but as arenas of battle and transport. When he explained his reasons for stationing troops in Greenland and Iceland, or sending war fleets to the South Pacific and the Indian Ocean, confused Americans reared on the Mercator projection had to consult other maps or globes to understand his logic. Thus alongside the proliferation of new and unusual maps in the 1940s we find endless discussions of the concepts of geography and mapping, which suggests that some Americans were becoming more sophisticated about the subtleties of cartographic design; in fact, during the 1940s there existed in the popular media a familiarity with maps that would have been inconceivable prior to the war.[33]

Helmuth Bay, assistant chief cartographer of Rand McNally, capitalized on geography's moment in the spotlight by giving lectures during and after the war. In 1943 Bay was invited to address an audience of publishers, authors, librarians, and readers at the New York Public Library to discuss the history and technique of mapmaking. That this topic had even been considered broadly interesting indicates what the war had done for geography. Bay acknowledged that librarians were noticing the inadequacy of their existing maps, and readers were "glibly requesting maps on a polar projection, on an equal area projection, or perhaps they come with an inquiry about laying out a great circle distance on a gnomonic chart."[34]

In 1942 *Life* devoted a feature to cartography, outlining each of the major projections, their relative advantages and disadvantages, and finally the need to improve on Mercator's limits. The article called the Mercator projection a "mental hazard in a war that is plotted on great circles across the land and sea and through the air." Harrison considered the *Life* article "a landmark in breaking down the mysteries of maps," explaining the rigidity of commercial mapmaking in a way that contributed to the public's acceptance of new, radically different images. Two months later the *New York Times Magazine* printed a similar feature—"Maps are Liars"—that graphically compared the different projections and discussed the inevitability of distortion in each. This subject was covered in dozens of articles in all kinds of magazines during the war, from *Reader's Digest* to the *American Scholar*, not to mention on the airwaves. The *New York Times* was particularly attuned to the nation's interest in geography, reporting on related matters such as Mayor LaGuardia's announcement in 1943 that the city's schools would reject the Mercator projection in favor of recent "globe" maps that better translated the realities of the air age. By 1944, two books addressed the problems of cartography, the first of their kind designed for a lay audience, and a number of geography departments were founded during the war.[35]

In its Christmas 1943 gift issue, *Consumer Reports* reviewed the market for maps, globes, and atlases, and endorsed the unconventional perspectives and projections popularized by Harrison while snubbing most of the more traditional maps. Of the globes recommended, half were titled either "polar view" or "air age." Many of these globes were not mounted on an axis, in the traditional way, but instead floated in a cradle so that viewers would be able to reimagine their sense of geography. Those readers not in the market for one of these new cradled globes were advised simply to take their own globe off its axis and place it in a bowl, thereby eroding strict—and therefore misleading—concepts of Northern,

Southern, Eastern, and Western hemispheres. By 1942 the *New York Times Magazine* was reporting on the reoriented maps and globes not as an element of politics or education, but as interior decoration, something no sophisticate ought to be without. War had made geography a consumer product.[36]

The number of articles, books, and radio shows devoted to mapping suggests an elevated awareness circulating through middlebrow and popular culture, and Rand McNally's own advertising campaigns reflect this changing sensibility. By 1937 most of the company's promotions were tied to political conflicts brewing around the world or to the anticipated arrival of the air age. Before the war one ad touted the arrival of aviation as revolutionary, as significant for the geographic imagination in the twentieth century as war and discovery had been in the past. An ad with a boy looking skyward and daydreaming of the adventures of flight referenced the enormously popular "Air Adventures of Jimmie Allen," a radio program centered around a teenager who flew around the world on dangerous missions. An equally explicit advertisement featured a globe centered on the North Pole that connected the cities of the world without reference to land or sea, since the relevance of these physical features as barriers to travel had been superseded by aviation. Yet while Rand McNally vocally promoted the relationship between war, the air age, and maps, its own products were only gradually revised to reflect those changes.[37]

This spatial reorientation was also fed by the heightened interest in geopolitical rhetoric discussed in chapter 6. Anxiety over Germany's rumored geopolitical sophistication even made the nation's continuing allegiance to the Mercator projection seem dangerous. The 1940s were rife with articles, maps, and textbooks calling for an overhaul of cartographic and geographic thought in order to adjust to changing political circumstances. One geopolitical treatise argued that the Mercator projection made the Atlantic and the Pacific a sort of "Maginot Line" around the Western Hemisphere that created in Americans a "psychological isolationism." In this and other ways the limits of the Mercator projection were exploited by geographers, cartographers, statesmen, and propagandists in order to promote the "science" of geopolitics. The geopolitical craze of the 1940s, premised on a conviction that power was linked to geography, was thus crucial to the spread of new maps during World War II. That the Germans were as close to Toledo and Abilene as they were to New York and Washington, and that the Japanese could attack Winnipeg as easily as San Francisco, did much to challenge standard renditions of the world on a map.[38]

THE TRANSFORMATION OF THE WORLD ATLAS

It is no coincidence that both Rand McNally and Hammond overhauled their world atlases after World War II. David Woodward argues that both companies saw the need to reconstruct their world atlases to adjust to postwar political changes, and Andrew McNally III recalled that during a postwar trip to Europe he recognized the need to adapt the atlas to the realities of the postwar world. But World War II was in fact far less disruptive to the map than World War I and its aftermath, when entire nations were created and others dissolved. A close look at the revised atlases of the postwar period reveals that the companies were responding primarily not to changing boundaries wrought by the war, but to changing conceptions of space wrought by the war and the air age, and to consumer pressures within the cartographic industry. The maps coming out of the National Geographic Society and newspapers had challenged cartographic conventions.

The introduction of an atlas has seldom been an occasion for media fanfare, but Rand McNally's *Cosmopolitan Atlas* was no ordinary atlas. A product of seven years of work by fifteen researchers, with a budget that totaled $450,000, not including printing costs, it was touted as the most expensive atlas in history, and Rand McNally's most comprehensive undertaking to date. Therefore, in the late 1940s the company engaged in an aggressive public relations campaign to promote the atlas as well as the company's reputation as a cartographic authority. The campaign was successful: the atlas debuted on October 22, 1949, and in just over a month nearly all copies of the first edition had been sold, not to schools and libraries but to individuals. Both within the publishing industry and in the mainstream media, Rand McNally found generous, unstinting praise for its effort.[39] In the company's own promotion of the atlas, the war and the air age were cited as causes for this total reinvention. Advertising copy made frequent reference to ideas made familiar during the war, such as scale, projection, and place-name spelling. Greater play was given to geopolitical concepts by mapping according to geographic region, history, and political association, rather than solely according to national borders.[40]

In fact, the *Cosmopolitan Atlas* did break conventions of both form and content that had been established over the company's half-century history. From 1927 to 1947 the company had made only minimal changes to the *Premier Edition* of their world atlas. In 1937 a few physical maps were introduced, as well as a few more depicting the American possessions abroad, but for the most part the interwar atlases contained only subtle variations. Thus when change did come it was all the more dramatic. Upon opening the 1949 edition of the *Cosmopolitan Atlas* the reader was instantly confronted with a different look and feel of the world on a map. Most noticeable was the appearance: Rand McNally had discarded wax engraving in favor of technologies advanced by the war, particularly photocomposition and offset lithography, which gave the map a much cleaner, more subtle, and sharper look. As David Woodward notes, offset printing technology had existed for some time, but its adoption was delayed by the vast amount of capital the major cartographic companies had invested in wax engraving equipment. Therefore, the switch to the new printing processes came slowly and idiosyncratically, but was hastened, I would argue, by the recent competition in the cartographic industry.[41] Compare this detail of Rand McNally's map of the state of Washington (fig. 9.12) with one from

Figure 9.12   Detail of a map created with Rand McNally's postwar use of offset printing. Compare with figure 8.7, an interwar map detail of the same region. Courtesy of the Harvard Map Collection.

the interwar period (see fig. 8.7). With these newer techniques the company was able to incorporate some topographic detail without sacrificing the overall clarity of appearance.

Just as striking as these printing changes was the arrangement of the maps: instead of being confronted first with maps of the United States and her territories, the reader found a more "global" and regionally ordered atlas. Foreign maps appeared first, organized according to regions rather than individual nations in order to convey physical and geopolitical relationships. This atlas depicted a world of political associations and political regions that had become markedly more pivotal to the United States. The atlas introduced maps of areas that previously had been paid only incidental attention, signaling a further increased awareness of political interdependence among nations. The standard Mercator map of the world, centered on the United States, was replaced with a map of the world on another cylindrical projection that partially corrected the former projection's distortions in the northern hemisphere, but eliminated Antarctica altogether. This map—"A Political Portrait of the World"—classified nations according to one of seven political categories. By bringing attention to areas of political unrest and movements for independence, the atlas turned the reader's eye toward political sovereignty and interdependence as the dominant organizational theme.[42]

This theme of interdependence was developed further through maps centering on the North Pole, emphasizing spatial proximity in the Northern Hemisphere. The first of these maps outlined flight routes and distances, while the second ominously suggested the proximity of North America to the Soviet Union, particularly suggestive because the latter surrounded nearly half of the Arctic Circle. Though polar maps had been included in prewar atlases, they were almost always used to illustrate routes of exploration, highlighting the physical, rather than the political, nature of the Pole (fig. 9.13). This was true in European atlases as well. In the highly respected and perennially successful German *Stieler* atlases, polar maps appeared as early as the turn of the century, yet the information on these maps was physical and topographical, not political. The date, distance, and route of Peary's exploration were central to these polar maps before the 1940s. The rise of aviation, together with the political urgency of World War II and the early cold war, created a different use for maps of the North Pole—illustrating the proximity of North America to Eurasia. The North Polar maps replaced the massive ocean buffers of the Atlantic and the Pacific on the Mercator projection with the relatively insubstantial Arctic, a "new mediterranean" surrounded by the two political superpowers

Figure 9.13 Typical interwar map of the North Pole, depicting the Pole as an arena of exploration only, and a relatively barren landscape. Courtesy of the Harvard Map Collection.

(fig. 9.14). Maps of Antarctica, if included at all, continued to reflect exploration routes and dates.

Equally powerful in suggesting geographic proximity rather than distance were the *Cosmopolitan Atlas* maps of the Atlantic and Pacific Oceans. By drawing the oceans on projections that attempted to correct for traditional distortions of the northern and southern latitudes, the company ended the emphasis on distance. The new map of the North Pacific linked the two continents at the Bering Straits and the Aleutian Islands, and modified the distance between Asia and North America. By identifying cities along the coasts and islands throughout the Pacific, the map depicted an ocean very different from the desolate expanse of the Mercator projection. These were oceans that linked, rather than separated, the land that bounded them (fig. 9.15).[43] In a similar way, Rand McNally's map of the Atlantic pulls the two hemispheres closer (fig. 9.16).

Overall, the maps of the first few editions of the *Cosmopolitan Atlas* were overwhelmingly political, but also incorporated some topographical detail. The company now mapped according to geopolitical regions rather than focusing exclusively on national borders, underlining the relationships *between* nations rather than their discrete identities. For example, Belgium, the Netherlands, and Luxembourg were now considered a natural unit, mapped together as the "Benelux nations." In mapping the United States as well, states considered to be natural geographic regions were mapped together rather than individually. Especially significant was the introduction of a map of Eurasia, a geographical region that had not been mapped as a unit since the nineteenth century, and which reflected the resurgence of Mackinder's heartland thesis.[44] The effect of the cartographic competition on the atlas was especially apparent in the look and style of the maps. With fewer place names and more subtle use of color, the company's maps looked remarkably like those of the National Geographic Society. President Andrew McNally III conceded that the company had been criticized for its reluctance to break from sure sellers, and that the *Cosmopolitan* was an attempt to address that criticism by emulating the European cartographic standard. But adherence to the standards set by the National Geographic Society and journalistic cartography was unmistakable as well. The atlas had begun to eliminate some railroads in western states and—like the Society—now downplayed county divisions, explaining that they were no longer an important mode of organization relative to the city, though arguably this was not the case for the American South.[45]

The company also changed the use of scale in its atlas, first by introducing and explaining to its readers the very concept of scale, which had not been attempted before in a popular atlas and which called attention to

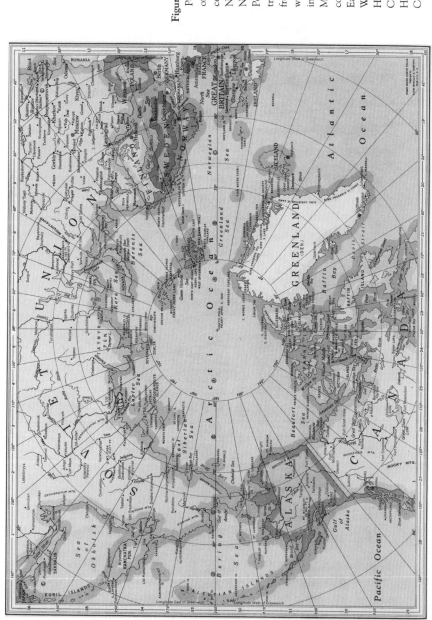

Figure 9.14

Postwar map of the world centered on the North Pole. Notice that the Pole has been transformed from a barren wasteland into a "new Mediterranean" connecting the Eastern and Western Hemispheres. Courtesy of the Harvard Map Collection.

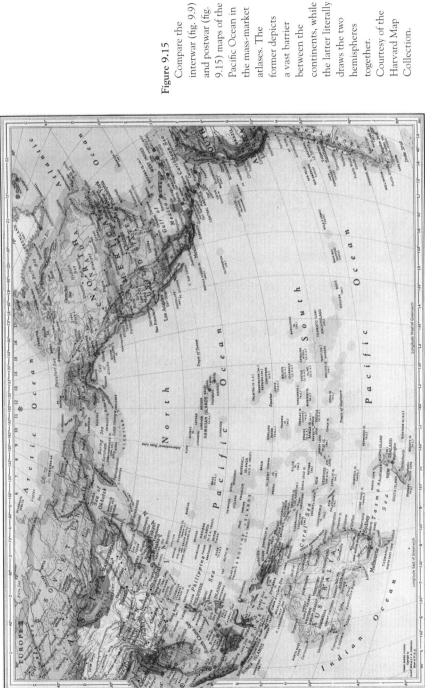

Figure 9.15

Compare the interwar (fig. 9.9) and postwar (fig. 9.15) maps of the Pacific Ocean in the mass-market atlases. The former depicts a vast barrier between the continents, while the latter literally draws the two hemispheres together. Courtesy of the Harvard Map Collection.

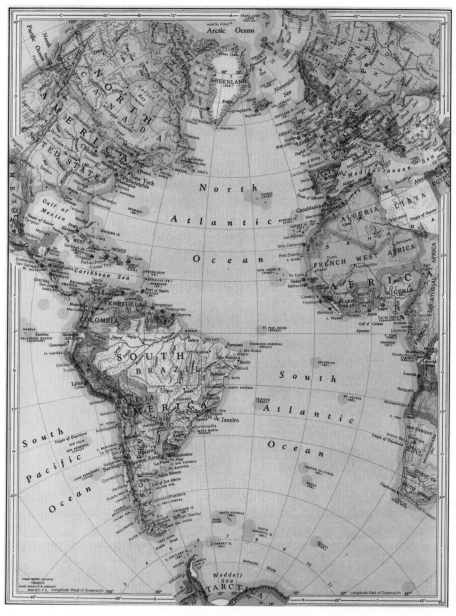

Figure 9.16 Rand McNally's postwar map of the Atlantic, a region that heretofore had not been given a separate map. Courtesy of the Harvard Map Collection.

the very enterprise of mapping as interpretive. By confronting the problem of scale, the company presented the cartographic process as fluid rather than static, themes that Harrison and countless others had emphasized during the war. This theme had been presented since the 1920s in Rand McNally's highly popular school atlas, but the *Cosmopolitan* marked the first time the company took such an approach in its popular atlases, a result of the vocal and persistent interest in the concept of mapping shown by the public and the media. What had been an idea appropriate only for students was now fit for popular consumption.[46] Rand McNally also took steps toward homogenizing the scale of its maps, which it had done previously only in atlases marketed to schools, so that readers could now "compare China with India or with Australia or with northern and southern Africa." Yet China could still not be compared with France, or Britain, or the United States, for their scales remained disproportionate. While China, India, Australia, northern Africa, and southern Africa were all mapped on a scale of one inch to 252 miles, France, the British Isles, and Germany were mapped as one inch to 63 miles; uniformity of scale was in this way approached, but not adopted completely. And though the scale of maps had been somewhat modified, the distribution of maps still heavily favored the United States. Nearly half the atlas was devoted to the United States, and another 13 percent to Europe, leaving Asia and Latin America with only 7 percent percent of the maps each, and Africa and Oceania with only 4 percent. These percentages constituted only a minor departure from previous atlases, indicating that in many regards the underlying message of the atlas had remained the same throughout the first half of the twentieth century.[47]

Considered as a whole, however, the *Cosmopolitan Atlas* capped a decade of remarkable change in which printing standards, cartographic perspectives, and atlas content had all been seriously reconceptualized. Up until 1939 Americans were dependent on a small community for the vast majority of their maps, and thus tended to encounter relatively homogenous products. While these cartographic companies were assured a market share throughout the war, their products were no longer the sole option for consumers. Creative and fresh approaches to mapping that were embraced by the media allowed the public to envision the world in radically different ways, which in turn encouraged a more sophisticated consumer market. As a result Rand McNally and Hammond had little choice but to reinvent their maps after the war to match the standards of their wartime competitors. The transformation of these maps and atlases accompanied a change in their very function and content, and the world encountered on a map

during the 1940s was fundamentally different from that of previous decades. World War II confirmed American stewardship abroad, underscoring
the emphasis on political interdependence in the atlases. The unmodified
Mercator projection was forever discredited, and the maps took a decisive
turn toward interdependence and spatial proximity. The ideological nature
of the war, the decline of imperial organization, and the birth of a bipolar
cold-war world all surfaced as well. By the end of the decade, the nation's
role as international steward was matched by a resolutely international cartographic self-portrait. These changes constituted a minor revolution in
the representation of the world on a map.

We frequently hear that ours is a world without borders, where geographical limits are easily overcome by technology, diplomacy, and the marketplace. People, ideas, images, and currency move effortlessly around the globe at a speed scarcely imaginable by previous generations. To live in a "global village"—whether real or anticipated—suggests that concepts of space and distance are malleable. Hopes of transcending geography also raise the obverse question: how was the world described, narrated, and envisioned at a time when most Americans could not experience it firsthand? Through geography, the vast and unintelligible was made discrete and coherent. Order was brought to chaos. Thus, from the 1880s to the 1940s, university and school geography, mass-market cartography, and the National Geographic Society strove to describe the world in a way that—above all—made *sense* to the public. At the turn of the century, professional geographers held out the hope that environmental explanations of human behavior would neatly unite the study of man and land within their discipline. The early twentieth-century *National Geographic* attempted to domesticate the exotic and to represent the world free from politics in an effort to deliver "mental relaxation without emotional stimulus." Cartographic firms anticipated the tastes of their audience by mapping the world in a way that appeared familiar, and therefore consumable.

This attention to the familiar—bred by repeated images and styles that eventually became conventions and traditions of their own—was a powerful influence over the history of geographical knowledge, and often took on a life of its own. In fact, geographical frameworks gained scientific legitimacy when they remained unchallenged over time and were thereby validated as natural, scientific, and unmediated representations of space. Simply put, maps, textbooks, and issues of the *National Geographic* accumulated power by virtue of their consistency. And in seeking to narrate the world as stable, orderly, and knowable, geography could become blind to very real changes occurring in the world. For instance, the map industry consciously chose to meet consumer expectations about the look and shape of the world by continuing to market the Mercator projection rather than diversifying its offerings. A few years later, when Franklin Roosevelt had to explain to confused citizens why he had stationed troops in Greenland and the South Pacific, he confronted a nation reared on the Mercator projection. To them these locations seemed illogical.

This example suggests that geographical knowledge—in its resistance to change and its effort to make sense of the world—has operated conservatively. But it has also functioned conservatively in its service to and defense of national power. Because mass-market cartography, academic and modern school geography, and the *National Geographic* each developed in an age of economic and territorial growth, they reflected and supported those imperatives. The early National Geographic Society simultaneously fed the federal government's expansionist imperatives and the public's curiosity about the world. Academic geography for many years drew upon intellectual frameworks that encouraged and legitimated expansion. Atlases and textbooks deployed categories of race, nation, and continent in order to represent a world that implicitly welcomed American intervention. In these and other ways the rise of American stewardship abroad influenced the development, definition, and in fact the very meaning of geography.

Neil Smith has pushed this relationship between geography and history even further. Smith argues that the advent of modern American liberalism in the twentieth century has quietly but forcefully guided our ability to apprehend the geographical dimension of American foreign policy. Liberal universalism, he argues, values the free flow of capital, and thus effectively erodes the explicitly spatial perspective of geography. Smith identifies in the American past a strain of what he calls geographical escapism, which has led to the assumption that we can reach "beyond geography." Referring to the liberal exceptionalist view of American history, he writes:

> The continental "isolation" of the US from Europe together with
> a different political tradition absents the US from the internecine
> power politics of global contest, freeing it to conduct foreign pol-
> icy on a moral rather than a narrowly pragmatic plane. Distance
> from Europe becomes a kind of founding myth of US foreign rela-
> tions; generally above power politics, US foreign policy is by the
> same token above the territorial contests that are the substance of
> this power politics. *Geography is deployed to justify its own erasure*.[1]

As Smith implies, this relationship between geography and history was reciprocal. Historical circumstances and imperatives influenced the way Americans were taught to understand the world. In turn, these geographical ideas envisioned what was actually *possible*.

This reciprocal relationship is complicated. I have suggested in broad terms how historical developments—foreign policy, technological advancements, educational reforms, and the like—have shaped geographical knowledge. But as I have also argued, geographical narratives frequently took on a life of their own, bred by intellectual conventions and habits developed over time or by perceptions about the expectations of the public. These conventions of geographical knowledge can in turn shape history. Thongchai Winichakul has nicely formulated this problem through the history of cartography in modern Indonesia. In an important passage quoted in Benedict Anderson's *Imagined Communities*, Winichakul writes:

> In terms of most communication theories and common sense, a
> map is a scientific abstraction of reality. A map merely represents
> something which already exists objectively "there." In the history
> I have described, this relationship was reversed. A map anticipated
> spatial reality, not vice versa. In other words, a map was a model
> for, rather than a model of, what it purported to represent. . . . It
> had become a real instrument to concretize projections on the
> earth's surface.[2]

Both Winichakul and Smith suggest that geographical knowledge has governed history in ways that have scarcely been noticed. This power of geography to shape history is difficult to apprehend, and even more difficult to document systematically. But the elusive and reciprocal nature of this phenomenon should not deter us from investigating its implications for the historical enterprise. We can never, of course, reach "beyond geography," for it is impossible to imagine the world outside of its interpretive conventions. But we can ask *how* geography has mediated the world for us, and how it has concretized the abstract. A tool as powerful as this—which has so often been taken at face value—deserves close scrutiny. I hope this study takes us a step in that direction.

This reciprocal relationship between geography and history is illustrated by the "war of the maps" in the 1940s. As explained in chapter 9, the decade witnessed a remarkable reorientation in geographic understanding, forced by the epic shock of World War II. Richard Edes Harrison—like Erwin Raisz and others—was able to transform the abstract and immense upheavals of the war into simple, tangible realities. Harrison and his contemporaries brought a strategic aerial view of the landscape to an industry where convention had historically ignored geopolitics and topography. In fact, the graphic perspectives advanced during the war were possible, powerful, and impossible to ignore precisely *because* they challenged conventional understandings that had reigned for generations. Most importantly, they managed to present the hidden geography of the war in a way that made intuitive *sense*. And yet such a wide-ranging reevaluation of geographic and cartographic conventions and traditions could not be sustained. However popular at the time, Harrison's perspectives were suited to the particular circumstances of the war, and are dated by that conflict. His perspective maps would return episodically — as in the missile crisis of the cold war, giving graphic pictorial emphasis to the threat of Soviet power — but would not endure. When news of a distant earthquake or a revolution broke, Americans sought out the *location*—not necessarily the geographic situation or larger context — of that place. By contrast, the Mercator-style projections have succeeded—despite their flaws—because of both their historical power and their narrative power to contain the world within a single view. These projections carried a kind of omniscient, totalizing quality that Harrison's views lacked.

Consider that in 1992 the Smithsonian held a short but popular exhibit on "The Power of Maps," designed to challenge the notion that maps are transparent representations of reality. The exhibit deconstructed dozens of maps of every kind, asking what information they included, what they concealed, and how they framed the landscape. The visitor entered the exhibit through a long hallway, surrounded by enlargements of Harrison's perspective maps taken from his *Look at the World* atlas, a huge seller upon its debut in 1944. This exhibit used Harrison's perspectives to challenge the standard "mental maps" we carry in our heads, and prompted among viewers a sense of disbelief that his creative and argumentative maps had remained dormant for nearly fifty years. Harrison completely reconfigured the "look of the world" in the 1940s, yet he is unknown today, a man whose work was resurrected briefly for the exhibit but who died in obscurity in the winter of 1994. Conversely, Gerhard Mercator's name has been preserved for centuries. Despite its limits, the basic tenets of the Mercator projection have become part of our geographic sensibility — a kind of metageography. However popular and powerful Harrison's perspectives were at mid-century, they could only modify — never displace — the basic coherence of the Mercator world. This is the world we recognize. This is the world that is familiar. History has powerfully shaped conventions and narratives of geographical knowledge in American culture. In turn, those narratives and conventions have — quite literally—enabled Americans to *see* and to make sense of the world.

Chapter One

1. Bruce Grant, quoting W. G. North in interview with Andrew McNally III, in "History of Rand McNally" (undated manuscript, c. 1956), 21. Rand McNally Collection, Newberry Library, Chicago.

2. Edward Said, *Orientalism* (New York: Vintage, 1978), chap. 1, pt. 2. Kären Wigen and Martin Lewis, *The Myth of Continents: A Critique of Metageography* (Berkeley: University of California Press, 1997), ix.

3. An excellent exception to this generalization is John Pickles, "Text, Hermeneutics, and Propaganda Maps," in *Writing Worlds: Discourse, Text, and Metaphor in the Representation of Landscape*, ed. Trevor Barnes and James Duncan (London: Routledge, 1992).

4. Denis Wood and John Fels, *The Power of Maps* (New York: Guilford Press, 1992), 21.

5. See J. B. Harley, "Maps, Knowledge, and Power," in *The Iconography of Landscape: Essays on the Symbolic Representation, Design and Use of Past Environments*, ed. Dennis Cosgrove and Stephen Daniels (Cambridge: Cambridge University Press, 1988); J. B. Harley and David Woodward, *The History of Cartography*, vols. 1, 2 bks. 1–3 (Chicago: University of Chicago Press, 1987–1998); Guntram Henrik Herb, *Under the Map of Germany: Nationalism and Propaganda, 1918–1945* (London: Routledge, 1997); Matthew H. Edney, *Mapping an Empire: The Geographical Construction of British India, 1765–1843* (Chicago: University of Chicago Press, 1997); Thongchai Winichakul, *Siam Mapped: A History of the Geo-Body of a Nation* (Honolulu: University of Hawaii Press, 1994).

6. J. B. Harley, "Deconstructing the Map," *Cartographica* 26, no. 2 (1989): 1–20, and esp. p. 13.

7. Gilbert H. Grosvenor, letter to David Fairchild, 10 June 1929, in Grosvenor Part II, Box 6 "NGM," Grosvenor Papers, Manuscript Division, Library of Congress.

8. Margaret Wilson, undated letter to the Society (c. 1926), in "Commendations and Criticisms, 1921–1926," Microfilm, Records Division, National Geographic Society Archives, Washington, D.C.

9. Jane Collins and Catherine Lutz, "Becoming America's Lens on the World: *National Geographic* in the Twentieth Century," *South Atlantic Quarterly* 91, no. 1 (winter 1992): 173; see also their *Reading National Geographic* (Chicago: University of Chicago Press, 1993).

10. Halford J. Mackinder, "On the Scope and Methods of Geography," *Proceedings of the Royal Geographical Society* 9 (1887): 141; Frederick Jackson Turner, *The Significance of the Frontier in American History* (New York: Holt, Rinehart, and Winston, 1962).

11. David N. Livingstone and J. A. Campbell, "Neo-Lamarckism and the Development of Geography in the United States and Great Britain," *Transactions of the Institute of British Geographers*, n.s., 8 (1983): 267–94. The research on the history of geography has grown impressively in the past two decades, particularly as it relates to problems of imperialism. Some of the more important works include: Brian Hudson, "The New Geography and the New Imperialism: 1870–1918," *Antipode* 9, no. 2 (1977): 12–19; Anne Godlewska and Neil Smith, *Geography and Empire* (Oxford: Basil Blackwell, 1994); and Felix Driver, "Geography's Empire: Histories of Geographical Knowledge," *Environment and Planning D: Society and Space* 10 (1992): 23–40.

12. Some scholars have recently investigated the complex history of professional geography and state power in more detail. See for example Neil Smith, *The Geographic Pivot of History* (Baltimore: Johns Hopkins University Press, forthcoming).

13. David N. Livingstone, "Climate's Moral Economy: Science, Race, and Place in Post-Darwinian British and American Geography," in *Geography and Empire*, ed. Godlewska and Smith, 154.

14. Classroom scene taken from Ellen Glasgow, *The Voice of the People* (New York: J. J. Little & Company, 1900), 44.

Chapter Two

1. On the gazetteer tradition, see Patricia Cline Cohen, "Statistics and the State: Changing Social Thought and the Emergence of a Quantitative Mentality in America, 1790 to 1820," *William and Mary Quarterly* 38, no. 1 (1981): 35–55; Daniel Calhoun, "Eyes for the Jacksonian World: William C. Woodbridge and Emma Willard," *Journal of the Early Republic* 4 (spring 1984): 1–26; Walter W. Ristow, "The French-Smith Map and Gazetteer of New York State," *Quarterly Journal of the Library of Congress* 36, no. 1 (winter 1979): 68–90.

2. Theodore Steinberg, "New England in a Pocketbook: Gazetteers and the Modernization of Landscape," *American Studies* 35, no. 2 (1994): 59–60, 67.

3. Walter W. Ristow, *American Maps and Mapmakers: Commercial Cartography in the Nineteenth Century* (Detroit: Wayne State University Press, 1985), chap. 25.

4. Martin Brückner, "Lessons in Geography: Maps, Spellers, and Other Grammars of Nationalism in the Early Republic," *American Quarterly* 51, no. 2 (June 1999): 311–43.

5. Jedediah Morse, *Geography Made Easy* (1800 edition), quoted in John A. Nietz, *Old Textbooks* (Pittsburgh: University of Pittsburgh Press, 1961), 200. Morse continued his interest in cartography by authoring atlases and gazetteers of the United States with his son Sidney Edwards Morse.

6. Benedict Anderson, *Imagined Communities: Reflections on the Origin and Spread of Nationalism*, rev. ed. (London and New York: Verso Press, 1991), 170–78.

7. Matthew H. Edney, "Cartographic Culture and Nationalism in the Early United States: Benjamin Vaughan and the Choice for a Prime Meridian, 1811," *Journal of Historical Geography* 20, no. 4 (1994): 385.

8. Ibid., 390.

9. Ibid., 386.

10. Arthur Robinson, "Mapmaking and Map Printing: The Evolution of a Working Relationship," in *Five Centuries of Map Printing,* ed. David Woodward (Chicago: University of Chicago Press, 1975), 5–6, 8, 14–15.

11. Henry S. Tanner's most respected and competitively priced atlases, distributed serially in the early 1830s, were still priced rather steeply at $15.00, well beyond the reach of artisans and laborers who earned daily wages of less than $2.00. The most successful among these was *Mitchell's New Universal Atlas,* printed through 1859, and *Colton's General Atlas,* printed through 1888. On pre–Civil War cartographers see Ristow, *American Maps.* On lithography see Erwin Raisz, "Outline of the History of American Cartography," *Isis* 26 (1937): 383; and W. W. Ristow, "Lithography and Maps, 1796–1850," in Woodward, *Five Centuries,* 106, 112.

12. David Bosse, *Civil War Newspaper Maps: A Cartobibliography of the Northern Daily Press* (Westport, Conn.: Greenwood Press, 1993), viii.

13. See William Goetzmann, *Army Exploration in the American West, 1803–1863* (New Haven: Yale University Press, 1959).

14. On the rivalries between the surveys, see John Noble Wilford, *The Mapmakers: The Story of the Great Pioneers in Cartography from Antiquity to the Space Age* (New York: Vintage, 1981), 213–14; and Matthew H. Edney, "Politics, Science, and Government Mapping Policy in the United States, 1800–1925," *American Cartographer* 13, no. 4 (1986): 295–306.

15. On book publishing and distribution in the late nineteenth-century West, see George Sereiko, "Chicago and its Book Trade, 1871–1893" (Ph.D. diss., Case Western Reserve University, 1973), esp. chaps. 1 and 3.

16. Michael Conzen, "Evolution of the Chicago Map Trade: An Introduction," in *Chicago Mapmakers: Essays on the Rise of the City's Map Trade* (Chicago: Chicago Historical Society, 1984), 10.

17. The introduction of wax engraved maps in 1872 brought map prices from the range of $5.00–10.00 to less than $1.00. Bruce Grant, "History of Rand McNally and Company" (undated manuscript, c. 1956), 21, Rand McNally Collection, Newberry Library, Chicago.

18. John Paul Goode, "The Map as a Record of Progress in Geography," *Annals of the Association of American Geographers* 27 (1927): 12; David Woodward, *The All-American Map: Wax Engraving and Its Influence on Cartography* (Chicago: University of Chicago Press, 1977), 24, 48–49; and Robinson, "Mapmaking."

19. This discussion of the cartographic industry and its techniques relies heavily on Woodward's *The All-American Map,* esp. 29, 39, 87–90.

20. Grant, "History," 8.

21. A *Business Tour of Chicago Depicting Fifty Years' Progress. Sights and Scenes in the Great City, Her Growing Industries and Commercial Development, Historical and De-*

scriptive. Prominent Places and People. . . . For Popular Distribution (Chicago: E. E. Barton, 1887).

22. Quote is from D. R. Cameron, of Cameron Amberg & Company, reprinted in "The 'Printer Laureate' of America," *Inland Printer* (June 1896). Since 1899 Rand McNally has been controlled and led by the McNally family. Andrew McNally headed the company until 1907, when he handed it to his son-in-law Harry Clow, who was president until 1933, when Andrew McNally II (grandson of Andrew McNally) assumed control. Andrew McNally III took over in 1948, and ran the company until 1974, when leadership was handed over to Andrew "Sandy" McNally IV. Andrew McNally III, interview with the author, Chicago, 10 June 1994; Andrew McNally III, *The World of Rand McNally* (New York: Newcomen Society of America, 1956).

23. Gerald A. Danzer, "George F. Cram and the American Perception of Space," in *Chicago Mapmakers*, ed. Conzen, 32–45.

24. Woodward, *All-American Map*, 45. Woodward explains well why wax engraving was never embraced by European mapmakers. Technically, copper engraving was far more entrenched in the European market by the time wax engraving made its debut. Furthermore, the aesthetic qualities of maps were of primary, not incidental, importance to European producers; wax engraving's typeface and cheap appearance had no place in their market.

25. Robinson, "Mapmaking," 22; Woodward, *All-American Map*, 124–27; Conzen, "Evolution," 10; see also Raisz, "History of Cartography," 383–86. On the nation's preoccupation with current and informational maps see Helmuth Bay, "The History and Technique of Map Making," *Eighth of the R. R. Bowker Memorial Lectures* (New York: New York Public Library, 1943), 13.

26. Woodward, *All-American Map*, 87–106.

27. Ads found in Unmarked Folder, Rand McNally Collection and from *Inland Printer* of January 1885. See also Cynthia Huggins Peters, "Rand McNally and Company—Printers, Publishers, Cartographers: A Study in Nineteenth-Century Mass Marketing" (M.L.S. diss., University of Chicago, 1981), 39.

28. *Youth's Companion* ad found in Unmarked Folder, Rand McNally Collection. The company began selling pocket maps of the states and territories in 1876, for about 50 cents to $1.00, which dropped to 25 cents by the 1890s, and 15 cents after that. By late 1912, the *Family Atlas* was advertised for $1.00. Grant, "History," 67.

29. This sentiment was also expressed by Andrew McNally III, interview with the author.

30. Woodward, *All-American Map*, 33–36.

31. Peters, "Rand McNally and Company," 67–68. In these cases, Rand McNally printed the newspaper's name on the cover and frontispiece of the atlas, such as the 1902 *Detroit Free Press* edition of the *New General Atlas of the World*, the 1885 *Chicago Weekly News* edition of the *New Household Atlas of the World*, or the *Unrivaled Atlas* of 1899, which was associated with the *Chicago Chronicle* and the *Philadelphia Public Ledger*. Rand McNally also arranged for "special editions" of these atlases to be associated with newspapers, such as the 1894 *New York Tribune* edition of the *Rand McNally Indexed Atlas of the World*, printed in eight parts, and heavily advertised in the paper. Grant, "History," 80–81.

32. Among the best selling atlases of the late nineteenth century, and the editions discussed here, are *The New Household Atlas of the World* (1885, 1898), *The New Indexed Atlas of the World* (1888), *Rand McNally's New Family Atlas of the World* (1888), *Rand McNally & Co.'s Pocket Atlas of the World* (1887–1939), *New Pictorial Atlas of the World* (1898), *New General Atlas of the World* (1895), *Rand McNally's Columbian Atlas of the World* (1898), *Cram's Unrivaled Family Atlas of the World* (1883), and *Cram's Unrivaled Atlas of the World, Indexed* (1890).

33. Johannes Dorflinger, "Austrian Atlases of the Late Nineteenth and Early Twentieth Centuries," in *Images of the World: The Atlas through History*, ed. John A. Wolter and Ronald E. Grim (New York: McGraw Hill, 1997), 233–56.

34. The standard of subcontinental coverage is taken from James Akerman, "On the Shoulders of a Titan: Viewing the World of the Past in Atlas Structure" (Ph.D. diss., Pennsylvania State University, 1991), chap. 5.

35. *Cram's Unrivaled Atlas* (1890), 9. The first American atlas to place the Western Hemisphere at the center, rather than on the left side, of the world map was Samuel Augustus Mitchell's *New Universal Atlas*, printed through 1859. This practice would not become common, however, until the twentieth century. See Alan K. Henrikson, "America's Changing Place in the World: From 'Periphery' to 'Centre'?" in *Centre and Periphery: Spatial Variation in Politics*, ed. Jean Gottman (Beverly Hills, Calif.: Sage Publications, 1980), 79–80.

36. See for example Hammond's *Modern Illustrated Atlas* of 1937. South America was also mapped regionally rather than according to national borders in *Philips' International Atlas* (1931).

37. Walter Ristow, *Maps for an Emerging Nation: Commercial Cartography in Nineteenth-Century America* (Washington, D.C.: Library of Congress, 1977), 27. Traditional geographic-topographic maps were increasingly accompanied by thematic maps in the late nineteenth century. These new maps widened the range of information shown in the atlas, including industrial development, transportation, foreign commerce, distribution of wealth, political opinion and party membership, and, occasionally, seasonal or climatic charts of the United States. Significantly, almost all the physical maps used in these late nineteenth-century atlases were climatic. On the rise of thematic mapping, see Wolfgang Scharfe, "German Atlas Development during the Nineteenth Century," in *Images of the World*, ed. Wolter and Grim; and Lester J. Cappon, "The Historical Map in American Atlases," *Annals of the Association of American Geographers* 69 (1979): 622–34.

38. Quote is from Rand McNally's *Pictorial Atlas* (c. 1898), 214. The importance of the textual information in these atlases has been somewhat contested. Some consider the text "filler" used to expand the size, and therefore the cost, of the atlases. Rand McNally generally used their own editors to write the accompanying textual information in the atlases, and Andrew McNally III considered it integral to the atlas as a whole. Because the textual descriptions constituted a sizable proportion of the total atlas throughout this period, and because they elaborated contemporary views of the world, they are here considered part of the overall mission of the atlas. Andrew McNally III, interview with the author, Chicago, 10 June 1994.

39. Rand McNally's *New Pictorial Atlas*, 218–19, 221–24, 227.

40. Ibid., 254–55.

41. Quote and illustration are from Rand McNally's *Pictorial Atlas of the World*, 165; see also Cram's *Unrivaled Family Atlas* (1883), 88.

42. As Lewis and Wigen have pointed out, the physical division between Europe and Asia is a tenuous and historically contingent one that is entrenched by the general insistence that the two regions form separate continents. Rarely are Europe and Asia mapped together. See Martin W. Lewis and Karen E. Wigen, *The Myth of Continents: A Critique of Metageography* (Berkeley: University of California Press, 1997), 21–31.

43. Cram's *Unrivaled Atlas* (1883), 92.

44. For example, the Rand McNally new *Standard War Atlas* was advertised for only 50 cents in the *Sioux City Tribune*. A special edition of this atlas was sold exclusively through newspapers, advertised in the *Tribune* as "well planned to answer the questions which people are asking about countries in distant parts of the world. The maps are in sufficient detail to be entirely intelligible and the low price at which the atlas is published, 50 cents, will make ignorance of the geography of the war unpardonable." *Sioux City Tribune*, 23 July [c. 1899].

45. Walter LaFeber, *The New Empire: An Interpretation of American Expansion, 1860–1898* (Ithaca: Cornell University Press, 1963).

46. Woodward, *All-American Map*, 48

47. *Rand McNally's War Atlas* (1898).

48. Rand McNally, *History of the Spanish-American War* (1898), 8–9, 12–13.

49. *Rand McNally's Atlas of Two Wars*, inset. Special attention was given to the trade and population of Manila, Iloilo, and Cebu.

50. *History of the Spanish-American War*, 9. The rich commercial potential of Puerto Rico and Hawaii was also emphasized in these atlases, in anticipation of future trade relations with the islands.

51. Ibid., 12–13. For instance, William Swinton's *Complete Course in Geography* (1875) did not include Alaska on any of its maps.

52. G. F. Cram & Company, *New Official Map of Alaska and the Klondike Gold Fields—The New 'Eldorado'* (1897), foldout map and 2.

53. See for example this map in *Rand McNally's Atlas of Two Wars* (1899).

54. Letter from A. F. Henning, Dallas, Texas, to the National Geographic Society, 5 February 1917, in "Magazine Suggestions: 1915–1923," Microfilm, Records Division, National Geographic Society, Washington, D.C.

55. J. B. Harley, "Deconstructing the Map," *Cartographica* 26, no. 2 (1989): 13.

Chapter Three

1. "The Annual Dinner of the National Geographic Society," *National Geographic Magazine* [hereafter NGM] 17, no. 1 (January 1906): 22–26.

2. The need for a kind of outlet among scientists was first addressed through the Cosmos Club, founded in 1878 by practicing scientists, including two who would later

be invited to the first meeting of the NGS—Grove Karl Gilbert and John Wesley Powell. In the wake of the Cosmos Club a number of more narrowly defined organizations sprang up, including the Biological Society in 1880, the Chemical and Entomological Society in 1884, the National Geographic Society in 1888, and the Geological Society of Washington in 1893. Of these, the National Geographic Society showed the greatest increase in membership. See James Kirkpatrick Flack, *Desideratum in Washington: The Intellectual Community in the Capital City, 1870–1900* (Cambridge, Mass.: Schenckman, 1975), 145.

3. Hubbard and Bell had also worked together on the magazine *Science,* which in 1900 became the organ of the American Association for the Advancement of Science. On Hubbard, see Flack, *Desideratum,* and Thomas Haskell, *The Emergence of Professional Social Science: The American Social Science Association and the Nineteenth-Century Crisis of Authority* (Urbana: University of Illinois Press, 1977). The five original founders sought out by Hubbard were Henry Gannett, chief geographer of the United States Geological Survey; A. W. Greely, leader of the Lady Franklin Bay Polar Expedition in 1882 and then for many years Chief Officer of the United States Army Signal Corps; A. H. Thompson, who worked for the United States Geodetic Survey; Henry Mitchell, a hydrographer with the USGS; and J. R. Bartlett, a hydrographer for the United States Hydrographic Office.

4. On the "active" versus "corresponding" classes of membership, see the Society's bylaws, in *NGM* 1, no. 2 (1889): 169–71. On lecture attendance, see Gardiner Greene Hubbard, "Synopsis of a Course of Lectures," in *NGM* 8, no. 1 (1897): 29. Hubbard's broad vision for the society is also reflected by the plan to distribute review copies of the *Geographic* to literary and scientific journals, geographers, geologists, meteorologists, and other foreign geographical societies. Letter from Henry Gannett to A. W. Greely, 7 November 1888, in Box 18, A. W. Greely Papers, Manuscript Division, Library of Congress.

5. W J McGee, "The Work of the National Geographic Society," *NGM* 7, no. 8 (August 1896): 257–58.

6. McGee, "Work of the National Geographic Society," 258.

7. Quote is from Grosvenor, *National Geographic Index,* vol. 1 (1961), 111. See also Bell, "The National Geographic Society," *NGM* 13, no. 3 (March 1912): 272–85. Grosvenor, who became editor of the *Geographic* in 1903 and president of the Society in 1920, wrote extensively on the history of the organization, thereby entrenching his interpretation and influencing other accounts. See Gilbert H. Grosvenor, "Alexander Graham Bell's Contribution to the National Geographic Society" (1922), in Box 79, Alexander Graham Bell Papers, Manuscript Division, Library of Congress [hereafter Bell Papers]; Grosvenor, *The National Geographic Society and Its Magazine* (Washington, D.C.: National Geographic Society, 1936, 1957); and Grosvenor, "The Romance of the Geographic," *NGM* 124, no. 10 (October 1963): 516–85. Howard S. Abramson's *National Geographic: Behind America's Lens on the World* (New York: Crown, 1987) relies entirely on Grosvenor's accounts for information on the Society's early years. Magazine histories also replicate this interpretation: see John Tebbel and Mary Ellen Zuckerman, *The Magazine in America, 1741–1990* (New York: Oxford University Press, 1991), 85–86; and Frank Luther Mott, *A History of American Magazines* (Cambridge: Harvard University Press, 1968 [1939]), 4.620–32.

8. Philip J. Pauly, "The World and All That Is In It: The National Geographic Society, 1898–1918," *American Quarterly* 31 (1979): 517–32.

9. *NGM* 16, no. 2 (February 1905): 87.

10. A. C. Fischer, "Oral History of Gilbert Hovey Grosvenor, 1962," in Records Division, National Geographic Society, Washington, D.C. (hereafter NGS Archives).

11. Haskell, *Emergence of Professional Social Science,* vii. There is also the odd case of John Hyde. Hyde served as editor of the *Geographic* and was part of the effort to remove Grosvenor and publish the magazine elsewhere in 1900, but never joined the "professional" AAG.

12. William Morris Davis, notes, in W.M. Davis file, Records Division, NGS Archives.

13. The most useful overviews of geography's early academic history include David N. Livingstone, *The Geographical Tradition: Essays in the History of a Contested Enterprise* (Oxford: Blackwell, 1992), chaps. 6–7; and Brian Blouet, *Origins of Academic Geography in the United States* (Hamden, Conn.: Archon Books, 1981).

14. John Hyde, "Introductory," *NGM* 7, no. 1 (January 1896): 2.

15. McGee, "Work of the National Geographic Society," 256.

16. Ibid., 256.

17. "Program for the Regular Course of Lectures of the National Geographic Society" (1897), pamphlet in Box 160, Grosvenor Papers; Hubbard, "Synopsis of a Course of Lectures," 32.

18. For instance, though the *Geographic* profiled Venezuela in 1896, this article was only tangentially related to the boundary crisis between that country and Britain. The important exception to this generalization is the substantial coverage of proposed canal projects in Panama and Nicaragua, which merited seven articles from 1889 to 1898, largely due to the sheer scale of the engineering efforts involved and their location in the Western Hemisphere.

19. O. P. Austin, "Our New Possessions and the Interest that They Are Exciting," *NGM* 11, no. 1 (January 1900): 32–33.

20. "Reception to Captain C. D. Sigsbee, U.S.N.," *NGM* 9, no. 5 (May 1898): 251–52.

21. Gannett, "Annexation Fever," *NGM* 8, no. 12 (December 1897): 355. Gannett spent 1903 in the Philippines, reporting his findings in "The Philippines and Their People," *NGM* 15, no. 3 (March 1904).

22. Robert T. Hill, "Cuba," *NGM* 9, no. 5 (May 1898): 241.

23. On Hilder see *Dictionary of American Biography,* p. x; and Robert Rydell, *All the World's a Fair: Visions of Empire at American International Expositions, 1876–1916* (Chicago: University of Chicago Press, 1984), 140–42.

24. After Hilder spoke on May 9, 1898 the Society's members also listened to a lecture describing the land and people of Puerto Rico. Hilder, "The Philippine Islands," *NGM* 9, no. 6 (June 1898): 282.

25. F. F. Hilder, "Gold in the Philippines," *NGM* 11, no. 12 (December 1900):

465–70; John Hyde, "Trade of the United States with Cuba," *NGM* 9, no. 5 (May 1898): 249.

26. John Hyde, "Commerce of the Philippine Islands," *NGM* 9, no. 6 (June 1898): 301–3; Charles E. Howe, "The Disposition of the Philippines," originally printed in *Financial Review* (27 May 1898).

27. Dean C. Worcester, "Notes on Some Primitive Philippine Tribes," *NGM* 9, no. 6 (June 1898): 284.

28. Austin, "Our New Possessions," 32.

29. Max Tornow, "Economic Condition of the Philippines," and Major A. Falkner von Sonnenburg, "Manila and the Philippines," *NGM* 10, no. 2 (February 1899); Sonnenburg quote on p. 72.

30. John Barrett, Late United States Minister to Siam, "The Philippine Islands and Their Environment," *NGM* 11, no. 1 (January 1900): 10–11.

31. Alexander Graham Bell, letter to Gilbert Hovey Grosvenor, 5 March 1900, in Part II, Box 6, file marked "NGM," Grosvenor Papers. Bell and Grosvenor also point to 1907 as an important year, when illustrations began to eclipse the text as the center of the magazine. Letters of Bell to Grosvenor, 17 October 1907 and 21 October 1907, in Box 99, Grosvenor Papers. The first hand-colored photographs were printed in William Chapin's "Glimpses of Korea and China," *NGM* 21, no. 11 (November 1910). Recently a number of scholars have investigated the relationship between geography and photography; see for example Joan M. Schwartz, "*The Geography Lesson:* Photographs and the Construction of Imaginative Geographies," *Journal of Historical Geography* 22, no. 1 (1996): 16–45; and James R. Ryan, *Picturing Empire: Photography and the Visualization of the British Empire* (London: Reaktion, 1997; Chicago: University of Chicago Press, 1997).

32. Rydell, *All the World's a Fair,* 140–42.

33. D. O. Noble Hoffman, "The Philippine Exhibit at the Pan-American Exposition," *NGM* 12, no. 3 (March 1901): 119–22. Significantly, the *Geographic* rarely referred directly to brutality and severity of Aguinaldo's uprising against the United States, which began in 1899. Only tacit references were made to this bloody conflict, and it never led the *Geographic* to question the legitimacy and urgency of America's presence. Aguinaldo, heralded as the leader of a liberation movement only months earlier, became a wily "opportunist" who had manipulated American goodwill in order to oust the Spanish from the islands. See Barrett, "The Philippine Islands and Their Environment," *NGM* 11, no. 1 (January 1900): 10–11.

34. Quoted in Rydell, *All the World's a Fair,* 160.

35. McGee's divided responsibilities between the Society and the Exposition caused no little concern that the former was drifting through 1904. See letter from Grove Karl Gilbert to W J McGee, 7 November 1904, Box 4, Folder "G" (1900–1905), McGee Papers, Library of Congress Manuscript Division.

36. Rydell, *All the World's a Fair,* 183; Colonel Clarence Edwards, "Governing the Philippine Islands," *NGM* 15, no. 7 (July 1904): 283–84.

37. Rydell, *All the World's a Fair,* 170–71; see also Eric Breitbart, *A World on Display: Photographs from the St. Louis World's Fair* (Albuquerque: University of New Mexico Press, 1997), esp. "The Philippine Reservation."

38. Catherine A. Lutz and Jane L. Collins, *Reading National Geographic* (Chicago: University of Chicago Press, 1993).

39. On the census in Cuba see NGM 11, no. 5 (May 1900): 205. The census was made available to *Geographic* readers for purchase in February 1901, and reports on the progress of efforts to contain yellow fever were printed in 1901 and 1902. On the Philippine census, see Gannett, "Philippine Islands," 98, 112.

40. Quote is from Gannett, "Philippine Islands," 107. On the *Geographic's* coverage of the general progress in the territories, see "American Development of the Philippines," NGM 14, no. 5 (May 1903); Colonel Clarence R. Edwards, "Governing the Philippine Islands," NGM 15, no. 7 (July 1904); "Progress in the Philippines," NGM 16, no. 3 (March 1905); Gilbert Grosvenor, "A Revelation of the Filipinos," NGM 16, no. 4 (April 1905); Secretary of War William Taft, "The Philippines," NGM 16, no. 8 (August 1905); Taft, "Some Recent Instances of National Altruism: The Efforts of the United State to Aid the Peoples of Cuba, Porto Rico, and the Philippines," NGM 18, no. 7 (July 1907); Taft, "Ten Years in the Philippines," NGM 19, no. 2 (February 1908).

41. W J McGee, address to the National Geographic Society, 28 March 1899, reprinted in NGM 10, no. 6 (June 1899): 202.

42. Ibid., 202; John Barrett, "The Philippine Islands and Their People," NGM 11, no. 1 (January 1900): 14.

43. John Hyde, "The National Geographic Society," NGM 10, no. 6 (June 1899): 222.

44. Ibid., 221; Preliminary Program of Lecture Course for 1899–1900, in Box 160, Grosvenor Papers.

45. Letter from Grosvenor to A. W. Greely, 10 July 1900, in Box 31, Greely Papers; letters from Bell to Grosvenor, 13 July, 15 August, and 16 August 1899, and from Grosvenor to Bell, 30 and 31 May 1899, all in Box 99, Grosvenor Papers.

46. The Philippines were covered in February, March, June, and July 1904; and in January, February, March, April, August, and November 1905. China was featured in 32 articles and three maps from 1900 to 1906; the Panama Canal in 25 articles and one map from 1889 to 1909 (16 after 1898); and Alaska 74 times—five times with maps—from 1898 to 1909.

47. Grosvenor, "Romance of the Geographic," 32, 55.

Chapter Four

1. Among the thirty-one founders of the AGS were at least six philanthropists and/or businessmen, about twelve publishers and editors, four historians, two clergymen, and one attorney. John K. Wright, *Geography in the Making: The American Geographical Society, 1851–1951* (New York: American Geographical Society, 1952), 14–18.

2. On Morse's negative influence, see Ernesto A. Ruiz, "Geography and Diplomacy: The American Geographical Society and the 'Geopolitical' Background of American Foreign Policy, 1848–1861" (Ph.D. diss, Northern Illinois University, 1975), 10–11, 25–29, 53; Jedediah Morse, *American Universal Geography, or, A View*

of the Present State of all the Empires, Kingdoms, States, and Republics of the Known World, and of the United States of America, in particular (originally published by Shepard Kollock, for the author, 1789). A good survey of the origins of government science can be found in A. Hunter Dupree, *Science in the Federal Government: A History of Policies and Activities* (Baltimore: Johns Hopkins University Press, 1986), chaps. 5 and 10.

3. David N. Livingstone, "Climate's Moral Economy: Science, Race, and Place in Post-Darwinian British and American Geography," in *Geography and Empire*, ed. Anne Godlewska and Neil Smith (Oxford: Blackwell, 1994), 154.

4. Mary Somerville, *Physical Geography*, 2d ed. (Philadelphia: Lea and Blanchard, 1850); George Perkins Marsh, *Man and Nature, or Physical Geography as Modified by Human Action* (New York: Charles Scribner, 1864). On Somerville and Marsh, see Geoffrey J. Martin and Preston E. James, *All Possible Worlds: A History of Geographical Ideas* (1972; New York: John Wiley, 1993).

5. Though Harvard was the first important source of geographical training in America, the University of California at Berkeley was the first major university to establish a separate department for geography, in 1898. Until that point, geography was usually found within departments of geology, or "geology and geography."

6. Halford J. Mackinder, "On the Scope and Methods of Geography," *Proceedings of the Royal Geographical Society* 9 (1887): 160.

7. Andrew Kirby, "The Great Desert of the American Mind: Concepts of Space and Time and Their Historiographic Implications," in *The Estate of Social Knowledge*, ed. JoAnne Brown and David van Keuren (Baltimore: Johns Hopkins University Press, 1991), 25.

8. In 1899 Davis was named the Sturgis Hooper Professor of Geology at Harvard, a position he retained until his retirement in 1912. On Davis's geographical ideas see R. J. Chorley, with R. P. Beckinsale and A. J. Dunn, *The History of the Study of Landforms, or the Development of Geomorphology. Volume 2: The Life and Work of William Morris Davis* (London: Methuen, 1973).

9. Davis, "Geography in the United States," *Science* 19 (1904): 120–32, 178–86.

10. Other geographical societies at the turn of the century include the Geographical Society of the Pacific (1881–1918), the Geographical Society of California (1891–1908), the Geographical Society of Chicago (founded in 1898), and the Geographical Society of Baltimore (1902–1912). See Gary S. Dunbar, "Credentialism and Careerism in American Geography, 1890–1915," in *Origins of Academic Geography in the United States*, ed. Brian Blouet (Hamden, Conn.: Archon Books, 1981).

11. Israel C. Russell, "An American Geographical Society," *Science*, n.s., 15, no. 370 (31 January 1902): 195.

12. Davis, "Geography in the United States," reprinted in *Scientific American Supplement* no. 1463 (16 January 1904): 23450–451.

13. Preston James and Geoffrey Martin, *The Association of American Geographers: The First 75 Years, 1904–1979* (Washington, D.C.: Association of American Geographers, 1978), 128.

14. Neil Smith, "Geography as Museum: Private History and Conservative Idealism in *The Nature of Geography*," in *Reflections on Richard Hartshorne's* The Nature of

Geography, ed. Neil Entrikin (Washington, D.C.: Association of American Geographers, 1989), 93.

15. By 1928, there were independent departments of geography at the University of California, University of Chicago, Clark University, University of Michigan, The Ohio State University, Ohio University (Athens), University of Wisconsin (Madison), and University of Minnesota. In universities at Nebraska, Missouri, Oklahoma, Pennsylvania, Tennessee, Wisconsin (Milwaukee), Cincinnati, Harvard, Northwestern, Washington (Missouri), Yale, Dartmouth, Oberlin, and Vassar, geography shared its department with another subject, usually geology or economics. Charles Redway Dryer, "A Century of Geographical Education," *Annals of the Association of American Geographers* 14 (1924): 147. Until 1935, most of the dissertations in American geography were written at Clark, Chicago, Wisconsin (Madison), Michigan, Pennsylvania, Cornell, and University of California, Berkeley. From 1935 to 1946, Harvard's production of graduate degrees rose. See Dean S. Rugg, "The Midwest as a Hearth Area in American Academic Geography," in Blouet, *Origins,* 189; Derwent Whittlesey, "Dissertations in Geography Accepted by Universities in the United States for the Degree of Ph.D. as of May, 1935," *Annals of the Association of American Geographers* 25 (1935): 211–37; and Leslie Hewes, "Dissertations in Geography Accepted by Universities in the United States and Canada for the Ph.D., June, 1935, to June, 1946, and Those Currently in Progress," *Annals of the Association of American Geographers* 36 (1946): 215–47.

16. A separate Department of Geography and Industry was founded at Pennsylvania in 1912. It was later moved to the School of Arts and Sciences and placed under the Department of Regional Science. The Department of Regional Science was permanently dismantled in 1993. Throughout its history at Pennsylvania geography had a distinctly commercial character.

17. Jurgen Herbst, "Social Darwinism and the History of American Geography," *Proceedings of the American Philosophical Society* 105 (1961): 538–44.

18. David R. Stoddart, "Darwin's Influence on the Development of Geography in the United States, 1859–1914," in Blouet, *Origins;* Stoddart, *On Geography and Its History* (Cambridge: Basil Blackwell, 1981), chap. 8 and esp. 167. See also Joanna E. Beck, "Environmental Determinism in Twentieth-Century American Geography: Reflections in the Professional Journals" (Ph.D. diss., University of California, Berkeley, 1985), 18–19.

19. George W. Stocking Jr., "Lamarckianism in American Social Science: 1890–1915," *Journal of the History of Ideas* 23 (1962): 239–40.

20. Herbert Spencer, *Principles of Psychology,* vol. 1 (1870), 422, quoted in Stocking, "Lamarckianism," 241–42.

21. David N. Livingstone and J. A. Campbell, "Neo-Lamarckism and the Development of Geography in the United States and Great Britain," *Transactions of the Institute of British Geographers,* n.s., 8 (1983): 267–94; Livingstone, "Evolution, Science and Society: Historical Reflections on the Geographical Experiment," *Geoforum* 16 (1985): 119–30.

22. Livingstone, *Geographical Tradition,* 222; and "Evolution, Science and Society," 124–26.

23. Central to Livingstone's argument about the influence of Lamarck is that Davis studied under and taught alongside Shaler, a lifelong advocate of neo-Lamarckian theory. Shaler himself was trained by Agassiz, whose opposition to Darwin encouraged Shaler's own receptivity to other evolutionary theories. Livingstone also connects Turner to Lamarck intellectually, arguing that the former drew heavily on the latter by focusing on the "generative and heritable influence of the environment" in shaping American history and national character. Campbell and Livingstone deepen this debt by discovering that Turner enthusiastically read Friedrich Ratzel's work in 1895 and 1896. Campbell and Livingstone, "Neo-Lamarckism," 271–73.

24. On the range of Lamarckian assumptions in geographical research, see Thomas F. Glick, "History and Philosophy of Geography," *Progress in Human Geography* 10 (1986): 273–74; and Livingstone, "Evolution, Science and Society," 125–29.

25. The literature on Ratzel is substantial; see especially Mark Bassin, "Imperialism and the Nation State in Friedrich Ratzel's Political Geography," *Progress in Human Geography* 11 (1987): 123–32; Alfred Thayer Mahan, *The Influence of Sea Power upon History, 1660–1783* (Boston: Little, Brown and Company, 1890).

26. Mackinder, "Scope and Methods"; Mackinder, "The Geographic Pivot of History," *Geographical Journal* 23 (1904): 421–37; Livingstone, "Evolution, Science and Society."

27. Kern, *Culture of Time and Space,* 223–24, 228.

28. Judith Conoyer Bronson, "Ellen Semple: Contributions to the History of American Geography" (Ph.D. diss., St. Louis University, 1973), 54–55, 80.

29. A clarification of terms here will be useful. Though few if any geographers used the term "environmental determinism" before World War I, when it was supposedly at the peak of its power, we will use it here to describe those who gave substantial weight to the influence of the *physical* elements—climate and topography especially—over human behavior. Throughout the 1920s, 1930s, and 1940s the term connoted physical rather than social surroundings.

30. Semple also saw a direct link between geographic conditions and political development, again drawing on the ideas of her teacher Ratzel.

"The struggle for existence is a struggle for space," says Ratzel. Abundant space in the United States has meant abundant opportunity and a chance for all to rise; it has developed in the Americans a powerful initiative and encouraged the democratic spirit. Thus as the isolation of North America helped the early colonists divest themselves of European monarchical principles of government, so the size of our country has kept classes and masses on a nearly equal footing by equality of opportunity.

From Semple, *American History and Its Geographic Conditions* (Boston: Houghton Mifflin, 1903), 227, 244.

31. Turner was enthusiastic about Semple's attempt to link the physical environment and human history, and endorsed more approaches like hers, which investigated "the influences of geography acting upon western expansion and shaping society to the resources of these vast provinces." Frederick Jackson Turner, "Geographical Interpretations of American History," *Journal of Geography* 4 (1905): 37.

32. Virginia M. Rowley, *J. Russell Smith: Geographer, Educator, and Conservationist* (Philadelphia: University of Pennsylvania Press, 1964), 21–23.

33. Along with Shaler, Davis trained many students who made important contributions to geography. Those students not discussed here include Curtis F. Marbut, who taught at Missouri from 1895 to 1910, and was from 1910 to 1935 in charge of soil surveys for the Department of Agriculture; Robert DeCourcey Ward, a climatologist at Harvard from 1890 to 1930; Lewis G. Westgate, who conducted surveys for the USGS; Alfred H. Brooks, later in charge of the USGS surveys of Alaska; and Douglas Johnson, a geologist at Columbia. See Martin and James, *All Possible Worlds*, 316–17.

34. Paul F. Griffin, "The Contribution of Richard Elwood Dodge to Educational Geography" (Ph.D. diss., Columbia University, 1952), 97–104.

35. Albert Perry Brigham, *Geographic Influences in American History* (Boston: Ginn and Company, 1903), 311, 314.

36. Brigham, *Geographic Influences*, 330–331.

37. The relationship between climate and human behavior was also of central importance to Ralph Tarr, who studied under Davis at Harvard until 1892 and then taught at Cornell until 1912. While there, Tarr compiled several texts with Frank McMurry that revolutionized school geography in the early grades. These texts, primarily the *Physical Geography* series of 1896, were the first school texts to humanize geography. Tarr headed Cornell's Department of Physical Geography in the 1890s — often as its only member — and worked to bring his department under the Department of Geology, which occurred only after his sudden death in 1914.

38. *The Pulse of Asia* (Boston: Houghton Mifflin, 1907, 1919, 1930). Huntington's work was encouraged in these years by Robert DeCourcey Ward, who was first one of his teachers at Harvard, then a colleague with similar ideas about the power of climate, and later a fellow member of the Immigration Restriction League in the early twenties, when Huntington's interest in eugenics flourished. See Martin and James, *All Possible Worlds*, chap. 11. For Huntington's statements on eugenics, see *The Character of Races: As Influenced by Physical Environment* (New York: Charles Scribner, 1924); Huntington and Leon F. Whitney, *The Builders of America* (New York: William Morrow, 1927); *Tomorrow's Children*, written in conjunction with the Directors of the American Eugenics Society (New York: J. Wiley and Sons, 1935); Huntington and Martha Ragsdale, *After Three Centuries: A Typical New England Family* (Baltimore: Williams and Wilkins, 1935); and Huntington, *Season of Birth: Its Relation to Human Abilities* (New York: J. Wiley, 1938).

39. Geoffrey Martin, *Ellsworth Huntington: His Life and Thought* (Hamden, Conn.: Archon Books, 1973), 64–69.

40. Ellsworth Huntington, *Civilization and Climate* (New Haven: Yale University Press, 1915), 3.

41. Ibid., 6.

42. Ellsworth Huntington, *World-Power and Evolution* (New Haven: Yale University Press, 1919), 17, 19.

43. Geoffrey Martin, Huntington's biographer, claims that Huntington was simply "not a Lamarckian," but has slim evidence to support his claim. Martin, *Ellsworth Huntington*, 243. Livingstone and Campbell, "Neo-Lamarckism."

44. Martin, *Ellsworth Huntington,* 114.

45. Huntington to H. E. Bourne, 7 November 1916, as quoted in Martin, *Ellsworth Huntington,* 142.

46. Geoffrey Martin, "The Emergence and Development of Geographic Thought in New England," *Economic Geography,* Special Issue (1998): 7.

47. Wright, *Geography in the Making,* 199–202.

48. Ellsworth Huntington, having worked for the Military Intelligence Division during the war, was invited to join the House Commission—later the Inquiry Committee—in November 1917, but declined for personal reasons. Mark Jefferson, however, accepted Bowman's offer to come aboard as chief cartographer to the Inquiry Committee. According to Charles Seymour, the Inquiry's work was essentially run by Bowman, Lippmann, and Jefferson. Martin, *Ellsworth Huntington,* 147; Geoffrey Martin, *Mark Jefferson: Geographer* (Ypsilanti: Eastern Michigan University Press, 1968), chap. 8; Charles Seymour, *Geography, Justice, and Politics at the Paris Conference of 1919* (New York: American Geographical Society, 1951); Lawrence Gelfand, *The Inquiry: American Preparations for Peace, 1917–1919* (New Haven: Yale University Press, 1963).

49. Geographers were also hired in other areas of government service during the war. See "Geographers in Pressing Demand for War Work," *Journal of Geography* 17 (1918): 33–34; "War Services of Members of the AAG," *Annals of the Association of American Geogrpahers* 9 (1919): 49–70; and Geoffrey Martin and Preston James, *All Possible Worlds: A History of Geographical Ideas* (1972; New York: John Wiley, 1993), 389–91.

50. George B. Roorbach, "The Trend of Modern Geography, A Symposium," *Bulletin of the American Geographical Society* 46 (November 1914): 798–808. Among the geographers polled were Isaiah Bowman, Albert Perry Brigham, Richard Elwood Dodge, Nevin Fenneman, Ellsworth Huntington, Mark Jefferson, Rollin Salisbury, Harlan H. Barrows, Charles R. Dryer, R. H. Whitbeck, and J. Russell Smith.

Chapter Five

1. William Swinton, *A Complete Course in Geography: Physical, Industrial, and Political* (New York: Ivison, Blakeman, & Taylor, 1875), 1.

2. Lawrence Cremin, *The Transformation of the Schools* (New York: Knopf, 1961), 17.

3. Joseph Mayer Rice, *The Public-School System of the United States* (New York: The Century Company, 1893), 59–60, 69–70, 139–41, 172–74.

4. Enrollment statistics taken from Henry Eldridge Bourne, *The Teaching of History and Civics in the Elementary and the Secondary Schools* (New York: Longmans Green & Company, 1900), 59.

5. Pennsylvania example from James Mulhern, *A History of Secondary Education in Pennsylvania* (Lancaster, Pa.: Science Press, 1933), 581. Illinois anecdote recounted in Charles Breasted, *Pioneer to the Past: The Story of James Henry Breasted, Archaeologist* (Chicago: University of Chicago Press, 1977 [1943]), 1–2. Examination for Admission, Jersey City High School, June 1885, reprinted in *The Wall Street Journal,* 9 June 1992.

6. The easy celebration of manifest destiny in nineteenth-century textbooks has been briefly explored in Laurence M. Hauptman, "Westward the Course of Empire: Geography Schoolbooks and Manifest Destiny, 1783–1893," *Historian* 40 (1978): 423–40; and Gerald Saxon, "How the West Was Taught," paper presented at the 1993 History of Cartography Meeting, Newberry Library, Chicago. Henry Miller Littlefield has treated this theme as well in a later context, in "Textbooks, Determinism, and Turner's Westward Movement in Secondary School History and Geography Textbooks, 1830–1960" (Ph.D. diss., Columbia University, 1967).

7. Samuel Augustus Mitchell, *A System of Modern Geography, Comprising a Description of the Present State of the World, and Its Grand Divisions, North America, South America, Europe, Asia, Africa, and Oceania, with Their Several Empires, Kingdoms, States, Territories, etc.* (Philadelphia: E. H. Butler & Company, 1873), 8.

8. Mitchell, *System of Modern Geography*. William Swinton's world schema was nearly identical to Mitchell's. See Swinton, *Complete Course in Geography*, 1–28.

9. Daniel Calhoun, "Eyes for the Jacksonian World: William C. Woodbridge and Emma Willard," *Journal of the Early Republic* 4 (spring 1984): 9–10, 15–17, 23.

10. Mitchell, *System of Modern Geography*, 235, 245, 296, 287, 290.

11. For studies that cover textbooks over the course of the nineteenth century, see Ruth Miller Elson, *Guardians of Tradition: American Schoolbooks of the Nineteenth Century* (Lincoln: University of Nebraska Press, 1964), esp. chap. 5; and John Nietz, *Old Textbooks* (Pittsburgh: University of Pittsburgh Press, 1961), chap. 6.

12. Mitchell, *System of Modern Geography*, 31.

13. Martin Lewis and Kären Wigen, *The Myth of Continents: A Critique of Metageography* (Berkeley: University of California Press, 1997), 43–44.

14. Mitchell, *System of Modern Geography* (1875 edition), 48.

15. Arnold Guyot, *Elementary Geography*. Some popular geography texts of the mid-nineteenth century include Matthew Fontaine Maury, *Physical Geography* (New York: Harper, 1868) and David Warren, *Common School Geography* and *System of Physical Geography* (Philadelphia: Cowperthwait, 1868–1879). Lewis and Wigen, *The Myth of Continents*, 30.

16. Jane Andrews, *Geographical Plays* (Boston: Ginn & Company, 1894 [1880]), introduction, 3, 24.

17. Frances Campbell Sparhawk, *Miss West's Class in Geography* (Boston: Lee & Shepard, 1887), 116.

18. Swinton, *Complete Course in Geography*, 17.

19. Occupational information from Hauptman, "Westward the Course of Empire," 438–40.

20. Quote from Sparhawk, *Miss West's Class*, 129.

21. National Education Association, *Report of the Committee on Secondary School Studies* (Washington, D.C.: Government Printing Office, 1893), 17 and passim. See also Lawrence Cremin, *A History of Education in American Culture* (New York: Holt, 1953), 389–92. Appointed in 1892, the Committee ultimately entrenched a rather traditional curriculum: Latin and Greek; English; French and German; Mathematics, Physics, Chemistry, and Astronomy; Natural History; History; and Geography. For a

survey of the Committee's work see Theodore Sizer, *Secondary Schools at the Turn of the Century* (New Haven: Yale University Press, 1964).

22. NEA, *Report on Secondary School Studies*, 204–5.

23. William Morris Davis, "Physical Geography in the High School," *School Review* 8 (1900): 388–404, 449–56; Ralph Tarr, "Physical Geography—The Teacher's Outfit," School Review 4 (1896): 161–72, 193–201.

24. "Report of the Committee on Physical Geography," *Addresses and Proceeding of the NEA*, vol. 38. Organized under NEA's Report of the Committee on College-Entrance Requirements (Chicago: University of Chicago, 1899), 783–87. Richard Hartshorne, "William Morris Davis—The Course of Development of His Concept of Geography," in *The Origins of Academic Geography in the United States*, ed. Brian Blouet (Hamden, Conn.: Archon, 1981), 142.

25. Alex Everett Frye, *A Complete Geography* (Boston: Ginn & Company, 1895); on the significance of this text in curriculum reform, see Katheryne T. Whittemore, "Celebrating 75 years of the *Journal of Geography*, 1897–1972," *Journal of Geography* 71 (1972): 7–18; Albert Perry Brigham and Richard Elwood Dodge, "Nineteenth Century Textbooks of Geography," in *The Teaching of Geography* (Bloomington, Ill.: NSSE, 1933); Charles Redway Dryer, "A Century of Geographical Education," *Annals of the Association of American Geography* 14 (1924): 125.

26. Jacques W. Redway and Russell Hinman, *Natural Advanced Geography* (New York: American Book Company, 1897), 3; Richard Elwood Dodge, *Reader in Physical Geography for Beginners* (New York: Longmans Green and Company, 1900); William Morris Davis, *Physical Geography* (Boston: Ginn and Company, 1898).

27. Statistics taken from Mulhern, *History of Secondary Education*, 548–49. Mulhern tabulated these figures from curriculum reports of principals gathered from state archives.

28. In 1890, 43,731 Americans graduated from high school in the United States; by 1900, that figure had more than doubled, and by the conclusion of World War I more than 300,000 students were completing their secondary education every year. Enrollment figures taken from the United States Department of Health, Education, and Welfare, *Digest of Educational Statistics*, Bulletin 1964, no. 18, 56, as reprinted in Littlefield, "Textbooks, Determinism, and Turner's Westward Movement," 391. On the professionalization of education at the turn of the century, see Cremin, *History of Education*, esp. chap. 11.

29. Frances FitzGerald, *America Revised: History Schoolbooks in the Twentieth Century* (Boston: Little, Brown and Company, 1979), 171. On educational trends in this period, see Cremin, *History of Education*, 440–41.

30. James F. Chamberlain, "Report of the Committee on Secondary School Geography," in *Addresses and Proceedings of the NEA* (Chicago: University of Chicago Press, 1909), 823; and Chamberlain, "Geography in the Life of the Pupil," *Proceedings of the NEA* (1907): 497–502. On the debate over the future of geography, see Ralph Tarr, Ray Whitbeck, Martha Krug-Genthe, and Mark Jefferson in "Results to Be Expected," *Journal of Geography* 4 (1905): 145–63.

31. Three of the six members of the committee were academics, and two of these—Mark Jefferson and Ray Whitbeck—had been trained in ontography, emphasizing the human response to physical environs.

32. Though many proponents of geography in the early twentieth century were insisting that physical geography remain the basis for school science, most saw the writing on the wall, and joined the call for reform. In the *Journal of Geography*, pedagogical articles after 1905 called for a stronger emphasis on the human response to the physical environment.

33. The report recommended that secondary schools offer one course on physical geography with a focus on the human response; a second course on commercial or industrial geography, regional geography, or human-physical geography; and a third course on the regional geography of the United States and western Europe.

34. FitzGerald, *America Revised*, 130–35.

35. The following texts are analyzed in this section: Jacques Redway, *Commercial Geography: A Book for High Schools* (New York: Charles Scribner's Sons, 1903); Albert Perry Brigham, *Commercial Geography* (Boston: Ginn and Co., 1911); J. Russell Smith, *Commerce and Industry* (New York: Henry Holt, 1925); Charles Redway Dryer, *High School Geography: Physical, Economic, and Regional* (New York: American Book Company, 1912); Spencer Trotter, *Geography of Commerce: A Textbook* (New York: Macmillan, 1911); Edward Van Dyke Robinson, *Commercial Geography* (Chicago: Rand McNally, 1922); and many of Frank G. Carpenter's *Geographical Readers* from the turn of the century.

36. Ralph S. Tarr and Frank McMurry, *A Complete Geography* (New York: Macmillan, 1904); Richard Elwood Dodge, *Elementary Geography: Part Two—Comparative Geography of the Continents* (Chicago: Rand McNally, 1904, 1907, 1916); Richard Elwood Dodge and William E. Grady, *World Relations and the Continents* (Chicago: Rand McNally, 1914, 1922); Richard Elwood Dodge and Earl Emmet Lackey, *Dodge-Lackey Elementary Geography* (Chicago: Rand McNally, 1927), and *Dodge-Lackey Advanced Geography* (Chicago: Rand McNally, 1928); Charles Redway Dryer, *High School Geography* (New York: American Book Company, 1912).

37. Brigham, *Commercial Geography*, iv, 59.

38. Ibid., 446–48.

39. Rand McNally, *Catalog of Maps, Globes, and Atlases for Schools* (Chicago: Rand McNally, 1912), 1–2.

40. Frank G. Carpenter, *Australia, Our Colonies, and other Islands of the Seas* (New York: American Book Company, 1904), 7, 161, 166, 201.

41. Carpenter, *Australia*, 357.

42. Tarr and McMurry, *Complete Geography*, 246.

43. Rand McNally, *Catalog of Maps, Globes, and Atlases*, 6, 32.

44. On the European mapping of Africa, see Thomas J. Bassett, "Cartography and Empire Building in Nineteenth-Century West Africa," *Geographical Review* 84, no. 3 (July 1994): 316–35.

45. Neil Harris has noticed a widespread American admiration for the Japanese at the turn of the century. Japanese exhibits at world's fairs were among the most prominent and popular of non-European participants from 1876 to 1904. This popularity was due, Harris argues, to the special tension in Japan between modernization and the preservation of ancient native traditions. As Harris puts it, "the Japanese were admired, paradoxically, both because they respected the West and because they re-

sisted it." Neil Harris, "All the World a Melting Pot? Japan at American Fairs, 1876–1904," in *Mutual Images: Essays in American-Japanese Relations*, ed. Akira Iriye (Cambridge: Harvard University Press, 1975), 24–54.

46. Carpenter, *Australia*, 346–47.

47. Tarr and McMurry, *Complete Geography*, 26, 27, 253.

48. Ibid., 264, 314, 317, 348; Carpenter, *Europe*, 11–12, 44, 86, 143–44, 186–87.

49. Florence Holbrook, *Rand McNally Elementary Geography* (Chicago: Rand McNally, 1901), 38–39, 42. Like other texts, this one voiced its firm support of the nation's recent growth, and included a relatively lengthy section on the new territories—and responsibilities—abroad (107–114).

50. Tarr and McMurry, *Complete Geography*, 464.

51. Edward Said, *Orientalism* (New York: Vintage Books, 1979), 71–72.

Chapter Six

1. Edward Yeomans, "Geography," *Atlantic Monthly* (February 1920): 167–69.

2. Ibid.

3. Randolph S. Bourne, *The Gary Schools* (1916; Cambridge: MIT Press, 1970), 248–49.

4. Ellwood P. Cubberley, *The Portland Survey: A Textbook on City School Administration Based on a Concrete Study* (Yonkers-on-Hudson, N.Y.: World Book Company, 1916), 136–53, 205–7, 260–63.

5. National Education Association, *The Social Studies in Secondary Education*, A Report of the Committee on Social Studies of the NEA on the Reorganization of Secondary Education, *Bulletin*, 1916, no. 28 (Washington, D.C.: Bureau of Education, 1916); A. David Hill and Lisa A. LaPrairie, "Geography in American Education," in *Geography in America*, ed. Gary L. Gaile and Cort J. Willmott (Columbus, Ohio: Merrill Publishing Company, 1989), 1–26.

6. U.S. Bureau of Education Bulletin no. 35, *Cardinal Principles of Secondary Education* (1918; Washington, D.C.: U.S. Government Printing Office, 1928), 2–3.

7. The National Council of Geography Teachers was founded in 1915 by George Miller, a normal school teacher in Minnesota, and included among its leaders Richard Elwood Dodge, Charles Redway Dryer, and Ray Whitbeck. Richard E. Dodge, "Some Problems in Geographic Education," *Annals of the Association of American Geography* 6 (1916): 13; George Miller, "National Council of Geographic Education," *Journal of Geography* 19 (1920): 69–76.

8. Begun in 1909, the journal became *Historical Outlook* in 1918, reaching some 5,000 history and social studies teachers, and potential NCSS members. In 1934 the journal's title was changed again, this time to *The Social Studies*, governed now by the American Historical Association. Thus, the official journal of the National Council of the Social Studies was housed under an explicitly historical organization. The *Journal of Geography* was founded in 1897, and originally run out of Columbia Teacher's College by Richard Elwood Dodge. In 1910 the *Journal* was transferred to Wisconsin, where it was run by Ray H. Whitbeck, a commercially minded geographer. In 1918

Isaiah Bowman sought to take control of the *Journal* through the American Geographical Society, but by March 1920 the AGS had decided to concentrate its efforts on research in order to strengthen its professional reputation. The *Journal* was therefore transferred to the NCGT, where it remains. See James W. Vining, *The National Council for Geographic Education: The First Seventy-Five Years and Beyond* (Indiana, Pa.: NCGE, 1990).

9. In 1920 311,000 seniors graduated from high school; by 1930 that figure had doubled; in 1940 it had doubled again to roughly 1.2 million. Actual enrollment figures were much higher, with 4.8 million attending high school in 1930, and 7.1 million ten years later. Statistics from Henry Miller Littlefield, "Textbooks, Determinism, and Turner's Westward Movement in Secondary School History and Geography Textbooks, 1830–1960" (Ph.D. diss., Columbia Univerity, 1967), 391.

10. Helen Goss Thomas, "The New Geography: Education for World Citizenship," *Educational Review* 59 (March 1920): 236–37.

11. Rose B. Clark, "Geography in the Schools of Europe," in Isaiah Bowman, *Geography in Relation to the Social Sciences* (New York: Charles Scribner's Sons, 1934), 329. For jeremiads about American geographic illiteracy, see Charles Redway Dryer, "A Century of Geographic Education," *Annals of the Association of American Geographers* 14 (1924): 123–24; Wallace W. Atwood, "The New Meaning of Geography in American Education," *School and Society* 13 (1921): 211–18; O. D. von Engeln, "A Campaign for Geography," *Journal of Geography* 18 (1919): 28–31.

12. Harlan H. Barrows, quoting an unnamed reformer, in "The Purpose of Geography Teaching," *Journal of Geography* 20 (1921): 151–54.

13. G. Stanley Hall, *Educational Problems* (New York: Appleton & Co., 1911), 556; and "The Ideal School," *Proceedings of the NEA* (1901): 480. On geography as a general science, see William Morris Davis, "Physical Geography in the High School," *School Review* 8 (1900): 388–404, 449–56; Ralph S. Tarr, "Physical Geography — The Teacher's Outfit," *School Review* 4 (1896): 161–72, 193–201; and T. H. Armstrong, "Relation of Geography to the Other Studies in the Elementary Course of Study," *Education* 23 (1903): 331–36.

14. "Physical Geography versus General Science," *School Science and Mathematics* 10 (1910): 761–72; "What Is Geography?" *Journal of Geography* 10 (1911): 59–60. On physical geography as the "mother of all sciences" see Miner H. Paddock, "Physical Geography in Our Public Schools," *Education* 25 (1904): 162; Harold W. Fairbanks, "Physiography: An Elementary Science Course in the High School," *Journal of Geography* 7 (1909): 217–26; Everett P. Carey, "General Science in Relation to Physical Geography," *Journal of Geography* 10 (1911): 62–66; P. E. Rowell, "General Science vs. Physical Geography," *School Science & Mathematics* 11 (1911): 116–21; Rollin G. Salisbury, "The Teaching of Geography — a Criticism and a Suggestion," *Journal of Geography* 8 (1909): 49–55; Salisbury, "Physiography in the High School," *Journal of Geography* 9 (1910): 57–63.

15. Daniel C. Knowlton, "The Relation of Geography to the Social Studies in the Curriculum," *Journal of Geography* 20 (1921): 226–27, 229.

16. Statistics taken from John Elbert Stout, *The Development of High-School Curricula in the North Central States from 1860 to 1918* (Chicago: University of Chicago, 1921), tables 29 and 30, 217–18 and tables A–J, 261–91; and W. M. Gregory,

"Secondary School Geography in the Middle West," *Journal of Geography* 8 (1910): 110–16.

17. New Jersey curriculum statistics from Henry W. Ridgway, "A Study of Representative Commercial Geography Textbooks, 1905–1925" (Ph.D. diss., University of Pennsylvania, 1930), 25–26.

18. Ray Hughes Whitbeck, with Loyal Durand Jr. and Joe Russell Whitaker, *The Working World* (New York: American Book Company, 1937), 37.

19. Leonard Packard and Charles Sinnott, *Nations as Neighbors: A Textbook in Geography for Junior High Schools* (New York: Macmillan, 1925), 14–15.

20. Leonard Packard, Charles Sinnott, and Bruce Overton, *The Nations at Work: An Industrial and Commercial Geography* (New York: Macmillan, 1933), v; Harlan H. Barrows and Edith Putnam Parker, *Countries Throughout the World* (New York: Silver Burdett, 1938).

21. *Goode Political and Physical Maps* (Chicago: Rand McNally, c. 1922), 5, 9. See also *Rand McNally Catalog of Maps and Globes for Schools* (Chicago: Rand McNally, 1922).

22. *A Catalog of Maps, Globes, and Atlases for Schools* (Chicago: Rand McNally, c. 1912), 83–87.

23. *Goode Political and Physical Maps*, 8.

24. Frank M. McMurry and A. E. Parkins, *World Geography: Book I, The New World* (New York: Macmillan and Company, 1927), 203–209; Wallace Atwood and Helen G. Thomas, *The Earth and Its Peoples: Nations Beyond the Seas* (Boston: Ginn and Company, 1930), 69.

25. Atwood and Thomas, *The Earth and Its Peoples*, 76, 47.

26. Edward Van Dyke Robinson, *Commercial Geography* (1915; Chicago: Rand McNally, 1922), 228.

27. Frank M. McMurry and A. E. Parkins, *World Geography: Book II, The Old World* (New York: Macmillan, 1925), 208–9.

28. Packard and Sinnott, *Nations as Neighbors*, 243.

29. *Rand McNally's Catalog of Educational Publications* (Chicago: Rand McNally and Company, 1928), 153, describing *Africa: A Geography Reader*.

30. Robinson, *Commercial Geography*, 344.

31. J. Russell Smith, *Commerce and Industry* (New York: Henry Holt, 1925), 540.

32. *Our Industrial World; Foreign Lands and Peoples; Human Use Geography; World Geography for Elementary Schools; Geography of the Americas;* and *Geography of Europe, Asia,* and *Africa* were among Smith's most popular texts, each reprinted throughout the 1930s and 1940s by the John C. Winston Company, Philadelphia. *Commerce and Industry* was printed by Henry Holt, intended for the secondary level, but also released for the college level under the title *Industrial and Commercial Geography*.

33. Another popular radio program was the "American School of the Air," which became required listening in schools across the country. Six states formally integrated the show into their curriculum, and an estimated six million students listened to the half-hour program every day, which for eighteen years told of geographic regions and

the feats of different American explorers. The show was the work of Lyman Bryson, a prominent educator at Columbia Teacher's College who was also a colleague of George T. Renner, a pivotal influence over the course of American school geography. During and after World War II, the imaginations of both were caught up in the potential of aviation, and especially its implications for geography. "New Horizons" and "Tales from Far and Near," recounted in John Dunning, *Tune in Yesterday: The Ultimate Encyclopedia of Old-Time Radio, 1925–1976* (Englewood Cliffs, N.J.: Prentice Hall, 1976), 28–29; Frank Buxton, *The Big Broadcast, 1920–1950* (1966; New York: Viking, 1972), 12; Jon Swartz, *Handbook of Old-Time Radio* (Metuchen, N.J.: Scarecrow, 1993), 81.

34. Joseph Corn, *Winged Gospel: America's Romance with Aviation, 1900–1950* (New York: Oxford University Press, 1983), chap. 6.

35. George T. Renner, *Human Geography for the Air Age* (New York: Macmillan, 1942); Corn, *Winged Gospel,* 127–29.

36. In some schools less decisive change was recorded. As of 1942, most of the geography courses in the white schools of Arkansas continued to be commercial, industrial, and economic, but the lack of change might have been due to the early date of the survey. James E. Collier, "Geography in the High Schools of Arkansas," *Journal of Geography* 42 (1943): 134–44; Clarence Burt Odell and Leslie Wood White, "The Status of Geography in the High Schools of Missouri," *Journal of Geography* 41 (1942): 41–51. Howard R. Anderson, "Offerings and Registrations in Social Studies," *Social Education* 14 (1950): 73–75.

37. Rex C. Miller, "High School Geography in Nebraska," *Journal of Geography* 47 (1948): 8–17; Dorothy Aldridge, "Present Status and Current Trends of Geographic Instruction in Massachusetts High Schools" (M.A. thesis, Clark University, 1948), introduction and 8–14, 26, 34–35; Ruby M. Junge, "Geography in the High Schools of Michigan," *Journal of Geography* 50 (1951): 329–34.

38. F. Webster McBryde, "Origin of the American Society for Professional Geographers: Take-Over and Cover-Up by Association of American Geographers Number 1," in *The American Society for Professional Geographers: Papers Presented on the Occasion of the Fiftieth Anniversary of Its Founding,* Occasional Publications of the AAG no.3 (April 1993): 6–7. At its height the ASPG had about a thousand members.

39. On this decision see Neil Smith, "Academic War over the Field of Geography: The Elimination of Geography at Harvard, 1947–1951," *Annals of the Association of American Geographers* 77 (1987): 155–72; and Thomas F. Glick, "Before the Revolution: Edward Ullman and the Crisis of Geography at Harvard, 1949–1950," paper presented at the AAG Annual Meeting, San Antonio, 26 April 1982

40. Gearóid Ó Tuathail, "'It's Smart to Be Geopolitical': Narrating German Geopolitics in United States Political Discourse, 1939–1943," in *Critical Geopolitics* (Minneapolis: University of Minnesota Press, 1996).

41. On the phenomenon of German *Geopolitik* see Eric Fischer, "A German Geographer Reviews German Geography," *Geographical Review* 38 (1948): 307–10; Thomas R. Smith and Lloyd D. Black, "German Geography: War Work and Present Status," *Geographical Review* 36 (1946): 398–408; L. K. D. Kristof, "The Origins and Evolutions of Geopolitics," *Journal of Conflict Resolution* 4 (1960): 15–51. More re-

cently attention has been paid to evaluating the myths and realities of this phenomenon. See for example Henning Heske, "German Geographical Research in the Nazi Period: A Content Analysis of the Major Geography Journals, 1925–1945," *Political Geography Quarterly* 5 (1986): 267–81; Henning Heske, "Karl Haushofer: His Role in German Geopolitics and in Nazi Politics," *Political Geography Quarterly* 6 (1987): 135–44; and Mark Bassin, "Race Contra Space: The Conflict between German Geopolitik and National Socialism," *Political Geography Quarterly* 6, no. 2 (1987): 115–34.

42. Halford J. Mackinder, *Democratic Ideas and Reality* (New York: Henry Holt, 1919), 150. See *Why We Fight,* especially Reel 2, "The Nazis Strike," National Archives, 1110F 1–7.

43. Halford J. Mackinder, "The Round World and the Winning of the Peace," *Foreign Affairs* 21 (July 1943): 595–605. Brian Blouet has made explicit connections between Mackinder's heartland thesis and the containment policies pursued in the early cold war. See Blouet, *Halford Mackinder, A Biography* (College Station: Texas A&M Press, 1987), esp. 365.

44. Attempts to "reclaim" the study of geopolitics from Haushofer can be found in Nicholas J. Spykman and Helen R. Nicholl, eds., *The Geography of the Peace* (New York: Harcourt Brace, 1944). Attempts to refute the association between German geopolitics and American political geography can be found in Hans W. Weigert and Vihjalmur Stefansson, *Compass of the World: A Symposium on Political Geography* (New York: Macmillan, 1944). For a contemporary review of this literature, see Bernard DeVoto, "The Falsity of Geopolitics in an Air Age," *New York Herald Tribune Weekly Book Review* 21, no. 5 (24 September 1944): 1–2.

45. George T. Renner, "Maps for a New World," *Collier's* (6 June 1942): 28; Renner, *Global Geography* (New York: Thomas Y. Crowell, 1944), 15.

46. George T. Renner, *Human Geography for the Air Age* (New York: Macmillan, 1942), 18, 42, 153; and Renner, "Theory of World Power and Control," in *Global Geography,* chap. 35.

47. Rand McNally Sales Catalog, "Air Education Adds a Third Dimension to Your Curriculum" (Chicago: Rand McNally, c. 1942), 1–2, Rand McNally Collection. "Rand McNally's Record in Maps and Globes," undated manuscript (c. 1947), Rand McNally Collection. See also "War Map of Europe," map advertisement for schools, and "Rand McNally Activities Maps and Globes for Social Studies Classes" (1940, Catalog #940), both uncatalogued, in Rand McNally Collection, Newberry Library, Chicago.

48. "Rand McNally School Maps, Globes, and Atlases" (Chicago: Rand McNally, 1942), and "Rand McNally's Air Globe" advertisement, uncatalogued, in Rand McNally Collection.

49. Polar maps are included in DeForest Stull and Roy W. Hatch, *Our World Today: Asia, Latin America, United States* (Boston: Allyn and Bacon, 1948); Jacob G. Meyer and O. Stuart Hamer, *The Old World and Its Gifts* (Chicago: Follet, 1942); Leonard Packard, with Bruce Overton and Ben Wood, *Our Air-Age World: A Textbook in Global Geography* (New York: Macmillan, 1944); Grace Croyle Hankins, *Our Global World: A Brief Geography for the Air Age* (New York: Gregg, 1944); James Franklin Chamberlain, *Air-Age Geography and Society,* revised by Harold E. Stewart (Chicago:

J. B. Lippincott Company, 1945); and Edith West, Dorothy Meredith, and Edgar B. Wesley, *Contemporary Problems Here and Abroad* (Boston: D. C. Heath and Company, 1947).

50. See for example Leonard Packard, with Bruce Overton and Ben Wood, *Geography of the World For High Schools* (New York: Macmillan, 1948); West, Meredith, and Wesley, *Contemporary Problems Here and Abroad*; Renner, *Human Geography for the Air Age*. Some, such as Packard, Overton, and Wood's *Our Air-Age World*; Whitbeck, Durand, and Whitaker's *The Working World*; Hankins's *Our Global World*; and Chamberlain's *Air Age Geography and Society*, devoted entire chapters to the Pacific.

51. Chamberlain, *Air-Age Geography and Society*, 645–65.

52. David A. Hollinger, *Postethnic America* (New York: Basic Books, 1995), 10 and chap. 3. Edward Steichen, *The Family of Man* (New York: Museum of Modern Art, 1955); Eric Sandeen, *Picturing an Exhibition* (Albuquerque: University of New Mexico Press, 1995).

53. Chamberlain, *Air-Age Geography and Society*, 665.

54. Meyer and Hamer, *The Old World and Its Gifts*.

55. Malvina Hoffman, "While Head Hunting for Sculpture," *New World Atlas*, 317.

56. "Description of the Races," *New World Atlas*, 318.

57. Meyer and Hamer, *The Old World and Its Gifts*, 21, 542.

58. Stull and Hatch, *Our World Today*, 34.

59. Chamberlain, *Air-Age Geography and Society*, 325–26.

Chapter Seven

1. William Morris Davis, letter to Henry Gannett, 22 October 1909, in W.M. Davis file, Records Division, National Geographic Society, Washington, D.C. (hereafter NGS Archives).

2. "The North Pole," *National Geographic Magazine* (hereafter NGM) (November 1909): 1008–9. Studies of the Pole controversy include Theon Wright, *The Big Nail: The Story of the Cook-Peary Feud* (New York: John Day, 1970); and John Edward Weems, *Race for the Pole* (New York: Henry Holt, 1960). There are also memoirs of the explorers themselves, including R. E. Peary, *The North Pole* (New York: Frederick A. Stokes, 1910); and F. A. Cook, *My Attainment of the Pole* (New York: Mitchell Kennerly, 1911).

3. Mrs. James R. Wilett, letter to the Society, 19 October 1916, in "Magazine Commendations and Criticisms, 1912–1921," Microfilm, Records Division, NGS Archives.

4. M. A. Gregory, Valparaiso, Indiana, letter to the Society, 14 March 1923, in "Magazine Commendations and Criticisms, 1921–1926," Microfilm, Records Division, NGS Archives.

5. Jane Collins and Catherine Lutz, "Becoming America's Lens on the World: *National Geographic* in the Twentieth Century," *South Atlantic Quarterly* 91, no. 1 (winter 1992): 163.

6. For a more extended treatment of Lutz and Collins's book-length study see Susan Schulten, "The Perils of *Reading National Geographic,*" *Reviews in American History* 23 (1995): 521–27.

7. Tamar Y. Rothenberg, *"National Geographic's* World: The Politics of Popular Geography, 1888–1945" (Ph.D. diss., Rutgers University, 1999), 12–14 [of copy in author's possession]. The term "strategy of innocence" is taken from Mary Louise Pratt, *Imperial Eyes: Travel Writing and Transculturation* (New York: Routledge, 1992).

8. For assumptions about the war fueling membership, see Rothenberg, *"National Geographic's* World," 30; Collins and Lutz, "Becoming America's Lens," 175. This same problem also limits an earlier survey of the *Geographic* written by Howard Abramson, who goes to great lengths to explore the Society's involvement in the well-known Peary-Cook controversy yet skims over the far more consequential experience of World War I. Howard S. Abramson, *National Geographic: Behind America's Lens on the World* (New York: Crown Publishers, 1987), 118–19.

9. Richard Ohmann, *Selling Culture: Magazines, Markets, and Class at the Turn of the Century* (London and New York: Verso, 1996).

10. Matthew Schneirov, *The Dream of a New Social Order: Popular Magazines in America, 1893–1914* (New York: Columbia University Press, 1994).

11. Ibid., 257.

12. Edwin C. Buxbaum, *Collector's Guide to the National Geographic* (Wilmington, Del.: Buxbaum, 1962), 234. In 1910 the magazine carried about 19 pages of ads per issue; by 1915 this number had risen to 39; in 1920 the average was 63, though hard economic times shrunk the number to 34 in 1921; by 1926 the average number of ad pages was up to 80. Despite the increasing proportion of advertising in the 1910s and early 1920s, the editors claimed that the *Geographic* remained largely independent of advertising receipts. Income from advertisements grew to about 20 percent of total receipts in the 1910s and 1920s. See Grosvenor, "Report of the Director of the National Geographic Society for the Year 1919," 4, in Box 160, Grosvenor Papers, Manuscript Division, Library of Congress, Washington, D.C. (hereafter Grosvenor Papers).

13. Gilbert Grosvenor, "Report of the Editor of the National Geographic Society, Gilbert Grosvenor for the Year 1921," in Box 100, Grosvenor Papers. The largest proportional increase in membership came in 1905, while the largest absolute increase came in 1960. From 1912 to 1916 membership rose by 400,000, and from 1922 to 1931 by about 500,000. Membership statistics taken from Memorabilia documents 91-7.17 through 91-7.1932, Records Division, NGS Archives.

14. Alexander Graham Bell, "Propositions by A. G. Bell," personal paper, 25 March 1903, in Box 254, Bell Papers. Letter from Bell to Grosvenor, 7 December 1905, in Box 99, Grosvenor Papers.

15. Margaret Wilson, National Committee of the YWCA, letter to the Society, 1926, in "Magazine Commendations and Criticisms, 1921–1926," Microfilm, Records Division, NGS Archives.

16. Alice Lindsay, letter to Society, 31 December 1922; and William A. Siemens, St. Louis, letter to the Society, 17 July 1921; both in "Magazine Commendations and Criticisms, 1921–1926," Microfilm, Records Division, NGS Archives.

17. William Warnholtz, Sioux City, Iowa, letter to the Society, 6 March 1923, in "Magazine Commendations and Criticisms, 1921–1926," Microfilm, Records Division, NGS Archives.

18. Lyndon Hansen, Farmington, Minnesota, letter to the Society, 19 September 1923, in "Magazine Commendations and Criticisms, 1921–1926," Microfilm, Records Division, NGS Archives.

19. Philip J. Pauly, "The World and All That Is In It: The National Geographic Society, 1898–1918," *American Quarterly* 31 (1979): 517–32, esp. 527.

20. "Visiting Your Society's Headquarters at Washington," Society pamphlet dated 1922, in folder marked "1925–1934," in Records Division, NGS Archives.

21. "The *Geographic*'s Upper Masses," pamphlet for advertisers dated 1930, in Records Division, NGS Archives.

22. Grosvenor, "Report of the Editor" (1921), 7–8, in Box 100, Grosvenor Papers.

23. Letter from Alexander Graham Bell to Gilbert Grosvenor, 13 July 1899, in Box 99, Grosvenor Papers.

24. The *McClure's* of 1900 was not yet the magazine of muckraking fame it would become after 1902.

25. On the *Century* and *McClure's* see Frank Luther Mott, *A History of American Magazines* (Cambridge: Harvard University Press, 1930–1968), vol. 3 chap. 21, and vol. 4 chap. 18.

26. Bell, "Address of the President of the National Geographic Society to the Board of Managers," 1 June 1900, 3, in Box 30, A. W. Greely Papers, Manuscript Division, Library of Congress, Washington, D.C. See also letters from Bell to Grosvenor, 30 May 1899 and 15 August 1899, in Box 99, Grosvenor Papers.

27. Letters from A. G. Bell to Grosvenor, 24 September 1899, 7 March 1901, 15 February 1902, 29 October 1902, 17 October 1907; and Bell's "Home Notes," 14 November 1907, all found in Box 99, Grosvenor Papers. See also Bell, "Home Notes," 21 November 1907, Box 100, Grosvenor Papers.

28. Letter from Bell to Grosvenor, 12 April 1904, 5–6 (emphasis in original), in Box 254, Bell Papers, Manuscript Division, Library of Congress, Washington, D.C. (hereafter Bell Papers).

29. Letter sent to prospective and current Society members, 1906, in Box 160, Grosvenor Papers.

30. Letter from A. G. Bell to Grosvenor, 5 March 1900, in Part II, Box 6, "NGM," Grosvenor Papers. Photograph appeared in Frank Frederick Hilder, "British South Africa and the Transvaal," *NGM* 11, no. 3 (March 1900): 94. This is the first example of photographs of barebreasted women in the *Geographic*. Abramson, as well as Collins and Lutz, mistakenly place the first instance of such photographs later, in 1903. See Abramson, *National Geographic*, 140–41, and Collins and Lutz, "Becoming America's Lens," 169.

31. William E. Curtis, "The Road to Bolivia," *NGM* 11, no. 6 (June 1900): 214. For Bell's vision of the illustrations in the magazine, see Bell, "Home Notes," 23 October 1907, copied by Charles R. Cox (Bell's secretary), in Box 99, Grosvenor Papers.

32. Willis Moore, Chief of the U.S. Weather Bureau, in "Our Climate Helps Us," *NGM* 15, no. 11 (November 1904): 452.

33. Colby Chester, "Haiti, A Degenerating Island: The Story of Its Past Grandeur and its Present Decay," *NGM* 19, no. 3 (March 1908): 206.

34. Theodore Roosevelt, "Wild Man and Wild Beast in Africa," *NGM* 22, no. 1 (January 1911): esp. 6.

35. Rothenberg, "*National Geographic*'s World," chap. 3, "Picturing Human Geography," esp. 124.

36. James Ryan, *Picturing Empire: Photography and the Visualization of the British Empire* (Chicago: University of Chicago Press, 1997), 61.

37. Eliza Scidmore, "Women and Children of the East," *NGM* 18, no. 4 (April 1907): 260; and "Madonnas of Many Lands," photographic essay in *NGM* 31, no. 6 (June 1917): 553. Others have noticed examples of this pattern elsewhere in the *Geographic*. Rothenberg, "*National Geographic*'s World," 157–58; and Lutz and Collins, *Reading National Geographic*, 79.

38. Grosvenor's high photographic standards were reaffirmed in the "Report of the Editor" (1921), 14, in Box 100, Grosvenor Papers.

39. Letter from Bell to Grosvenor, 21 October 1907, in Box 99, Grosvenor Papers

40. The Andes expeditions were highly popular in the *Geographic* during the 1910s, and the entire April 1913 issue was devoted to Peru. See Hiram Bingham, "Explorations in Peru," *NGM* 23, no. 4 (April 1912): 417–22; Bingham, "In the Wonderland of Peru," *NGM* 24, no. 4 (April 1913): 387–573; Bingham, "The Story of Machu Picchu: The Peruvian Expeditions of the National Geographic Society and Yale University," *NGM* 27, no. 2 (February 1915): 172–217; "Further Explorations in the Land of the Incas," *NGM* 29, no. 5 (May 1916): 431–73.

41. Letter from Gannett to Grosvenor, 6 January 1912, GHG file 11-10015.208, Records Division, NGS Archives. Grosvenor and Gannett also had plans for the Society to design an atlas; see letter from Grosvenor to Henry Gannett, 25 May 1912, GHG file 11-10015.208, Records Division, NGS Archives.

42. Grosvenor, "Minutes of Board of Managers," 20 January 1909, in Box 99, Grosvenor Papers; Grosvenor, "Notes Regarding the Bulletin of the American Geographical Society of New York," 2 January 1912, in Box 160, Grosvenor Papers.

43. Letter from Bell to Grosvenor, 13 August 1914, in Box 99, Grosvenor Papers.

44. "Geographic Editor Charmed with Exposition," *San Diego Union*, 12 July 1915, in Box 79, Bell Papers.

45. Grosvenor, "Report of the Director and Editor to the National Geographic Society for the Year 1914," *NGM* 27, no. 3 (March 1915): 319.

46. Pauly, "'The World and All That Is In It,'" 528.

47. "Wards of the United States: Notes on What Our Country is Doing for Santo Domingo, Nicaragua, and Haiti," *NGM* 30, no. 2 (August 1916): 143–84, quotes on 147, 155.

48. William Howard Taft, "Washington: Its Beginning, Its Growth, and Its Future," *NGM* 27, no. 3 (March 1915): 221–92.

49. Pauly, "'The World and All That Is In It,'" 528. Grosvenor's rather insistent buoyancy seemed incongruous in the penultimate photograph of this series, that of a young coal miner just emerged from his work, described as "decidedly in his element and supremely happy." See "Little Citizens of the World," *NGM* 31, no. 2 (February 1917): 148–63.

50. Richard Conniff, "Our Foreign-Born Citizens," *NGM* 31, no. 2 (February 1917): 95–130.

51. Sydney Brooks, "What Great Britain Is Doing," *NGM* 31, no. 3 (March 1917): 193–210, quote on 210.

52. Senator John Sharp Williams, "The Ties that Bind: Our Natural Sympathy with English Traditions, the French Republic, and the Russian Outburst for Liberty," *NGM* 31, no. 3 (March 1917): 281–86, quote on 286.

53. Quoted comments from readers all found in "Magazine Commendations and Criticisms, 1912–1921," Microfilm, Records Division, NGS Archives. Specific comments cited: A. E. Brown, letter, 7 May 1917; Herman S. Riedever, St. Louis, letter, 17 May 1917; Charles Todd, St. Louis, letter, 28 May 1917; H. Haessler, Milwaukee, letter, 25 June 1917; Anonymous (German newspaper editor), letter, 7 June 1917.

54. Robert Lansing, "Prussianism," *NGM* 33, no. 6 (June 1918): 546–57.

55. William G. Hern, Milwaukee, Wisconsin, letter, 18 November 1919, in "Magazine Commendations and Criticisms, 1912–1921," Microfilm, Records Division, NGS Archives.

56. Melville Chater, "The Land of the Stalking Death: A Journey through Starving Armenia on an American Relief Train," *NGM* 36, no. 5 (November 1919): 393–420, quote on 405.

57. Anonymous letter, 24 February 1917, in "Magazine Commendations and Criticisms, 1912–1921," Microfilm, Records Division, NGS Archives.

58. Letter from R. M. Mueller (Evangelical pastor), Hutchinson, Minnesota, to the Society, 2 March 1920; unauthored comment from 1917, both in "Magazine Commendations and Criticisms, 1912–1921," Microfilm, Records Division, NGS Archives.

59. As Grosvenor wrote to Bell in 1921, this doctrine would not only lead to the end of the British Empire, but "would justify the establishment of a black republic in the United States, and it seems to me would also justify the secession of our Southern States. The doctrine has been used to incite rebellions in Ireland, Egypt and India and I have felt [it] is one of the principal reasons of the present turmoil throughout the world." Letter from Grosvenor to Alexander Graham Bell, 13 May 1921, in Box 100, Bell Papers, quoted in Rothenberg, "*National Geographic*'s World," 104. William Howard Taft, "The League of Nations: What It Means and Why It Must Be," *NGM* 35, no. 1 (January 1919): 43–66.

60. Washington, D.C. *Evening Star*, 1 April 1919, in Box 160, Grosvenor Papers; telegram from Grosvenor to Bell, 21 January 1920, in Box 100, Grosvenor Papers.

61. Grosvenor, "Report of the Editor" (1921), 16, in Box 100, Grosvenor Papers.

62. Letter from Grosvenor to Bell, 23 April 1920, in Box 100, Grosvenor Papers.

63. William Morris Davis, "The Progress of Geography in the United States," *Annals of the Association of American Geographers* 14 (1924): 177.

64. *Elmhurst Press*, 29 March 1926, in "Magazine Commendations and Criticisms, 1921–1926," Microfilm, Records Division, NGS Archives.

65. Grosvenor, "Report of the Editor" (1921), in Box 100, Grosvenor Papers.

66. Edgar Herr Levan, "The Scribbler," printed in Lancaster (Pa.) *Examiner*, 29 June 1921 (copyright New Era Publishing Company), in "Magazine Commendations and Criticisms, 1912–1921," Microfilm, Records Division, NGS Archives.

67. Anonymous letter to the Society, 10 March 1925, in "Magazine Commendations and Criticisms, 1921–1926," Microfilm, Records Division, NGS Archives.

68. Letter from E. Murray, Ontario, to the Society, 28 January 1925, in "Magazine Commendations and Criticisms, 1921–1926," Microfilm, Records Division, NGS Archives. Murray might have been referring to the piece by A. M. Hassanein Bey, "Crossing the Untraversed Libyan Desert," *NGM* 46, no. 3 (September 1924).

69. Letter from Robert Harris, Pomfret, Connecticut, to the Society, 14 April 1924, in "Magazine Commendations and Criticisms, 1921–1926," Microfilm, Records Division, NGS Archives.

70. Letter from Seaborn Noxon, Westfield, New York, to the Society, 1926, in "Magazine Commendations and Criticisms, 1921–1926," Microfilm, Records Division, NGS Archives. Georges-Marie Haardt, "Through the Deserts and Jungles of Africa by Motor," *NGM* 49, no. 6 (June 1926): 651–720.

71. Elizabeth Bishop, "In the Waiting Room," in *The Complete Poems: 1927–1979* (New York: Farrar, Straus and Giroux, 1988), 159–62.

72. Letter from Rev. Harold MacIlvaine Dorrell, Lambertville, New Jersey, to the Society, 3 March 1923, in "Magazine Commendations and Criticisms, 1921–1926," Microfilm, Records Division, NGS Archives.

73. Letter from Grosvenor to Dr. David Fairchild, 10 June 1929, in Grosvenor Part II, Box 6 "NGM," Grosvenor Papers; emphasis added.

74. Douglas Chandler, "Changing Berlin," *NGM* 71, no. 2 (February 1937): 131–77; John Patric, "Imperial Rome Reborn," 31, no. 3 (March 1937): 269–325. See also Ruth Q. McBride, "Turbulent Spain," *NGM* 70, no. 4 (October 1936): 397–427.

75. Abramson, *National Geographic*, 169.

76. Letter from E. M. DeShar, Albany, N.Y., 30 September 1921, in "Magazine Commendations and Criticisms, 1912–1921," Microfilm, Records Division, NGS Archives.

77. Grosvenor, unpublished recollections (1940s), Box 160, Grosvenor Papers. Gilbert H. Grosvenor, *The National Geographic and Its Magazine* (Washington, D.C.: National Geographic Society, 1957), 5.

Chapter Eight

1. Cyrus C. Adams, "Maps and Mapmaking," *Bulletin of the American Geographical Society* 44 (1912): 198–201 and passim.

2. Dorflinger's study of Austrian atlases indicates that this trend was mirrored in Europe. The number of non-European regional maps in these atlases rose from about 20 percent in the 1870s to 30percent by World War I, corresponding to a decrease in the number of Austro-Hungarian and European maps. In these atlases, the United States was the first non-European area to be mapped with more detail, followed by east Asia, particularly China and Japan. Johannes Dorflinger, "Austrian Atlases of the Late Nineteenth and Early Twentieth Centuries," in *Images of the World: The Atlas through History*, ed. John A. Wolter and Ronald E. Grim (New York: McGraw Hill, 1997), 244–46.

3. Quote is from the *Chicago Inter Ocean* review of Rand McNally's *Business Atlas*, 6 March 1899, found in II Cartographic Publishing, Box 1, Rand McNally Collection, Newberry Library, Chicago.

4. See Rand McNally's *Library Atlas* (1912) and *Ideal Atlas* (1915). Interestingly, none of the African maps covered either the West African coast or the Congo region. Rand McNally's *New Family Atlas of the World* (1914) also devoted a large map just to Luzon, though by 1916 the Philippines were no longer mapped together with the United States, having been grouped instead with the other Pacific Islands in the *Imperial Atlas*.

5. Quote is from Alfred Sidney Johnson, an employee in the Map Department of Rand McNally, in his "How Maps and Atlases Are Made," *Publishers' Weekly* 101 (1922): 1166.

6. *Rand McNally & Co.'s Pocket Atlas of the World* (1887), 38, and from the 1900 edition, 331, 334, 339.

7. *Pocket Atlas* (1887), 179; *Pocket Atlas* (1900), 219, 222–23.

8. *Pocket Atlas* (1900), 222, 228; *Pocket Atlas* (1936), 246. Another example was the changing description of the soil in Cuba. In 1900 the atlas enthusiastically praised the soil's inexhaustible fertility; twelve years later it was considered only "highly favorable."

9. Rand McNally produced eight atlases geared to the war.

10. Rand McNally, *Graphic Representation of the Battle Fields of Today* (1915); Rand McNally's *Atlas of the World War* (1918); rival companies produced similar atlases, such as Hammond's *New Map of Europe, Showing Seat of Austro-Servian War* (1914), and Cram's *Atlas of the War in Europe* (1915).

11. An exception to the rule of small-scale maps in American atlases was C. S. Hammond's *Brentano's Record Atlas* (New York: C. S. Hammond, 1918). With maps of the western front drawn on a scale of 1:10 (1 inch to 10 miles), the Hammond atlas was able to illustrate political boundaries, railways, altitudes, wireless stations, fortresses, fortified towns, arsenals, aircraft depots, forests and woods, and canals.

12. "The National Geographic War-Zone Map," *National Geographic Magazine* 33, no. 5 (May 1918): 494.

13. Letter from J. R. Purser, Charlotte, N.C., to National Geographic Society, 14 September 1918, in "Suggestions—Film 1915–1923," Records Division, National Geographic Society.

14. Memo from Albert Holt Bumstead to Gilbert Grosvenor, 24 December 1915; item 11-10015.837, Records Division, National Geographic Society.

15. Letter from A. F. Henning, Dallas, Texas, to National Geographic Society, 5 February 1917, in "Suggestions—Film 1915–1923," Records Division, National Geographic Society. Caleb D. Hammond, interview with the author, 3 March 1995, Maplewood, N.J.

16. Rand McNally, *Atlas of the European Conflict* (1914).

17. Rand McNally, *New Reference Atlas of the World and the War* (1917). All quotes from 6, 10. See also *Atlas of the World War* (1918) and *Atlas of Reconstruction for Schools* (1921).

18. On radio broadcasts, see John Dunning, *Tune in Yesterday: The Ultimate Encyclopedia of Old-Time Radio, 1925–1976* (Englewood Cliffs, N.J.: Prentice Hall, 1976), 28–29; and Frank Buxton, *The Big Broadcast, 1920–1950* (1966; New York: Viking, 1972), 12. On the relationship between commercial mapping and automobile culture, see James Akerman, "Selling Maps, Selling Highways: Rand McNally's 'Blazed Trails' Program," *Imago Mundi* 45 (1993): 77–89; and Akerman, "Blazing a Well Worn Path: Cartographic Commercialism, Highway Promotion, and Auto Tourism in the United States, 1880–1930," *Cartographica* 30, no. 1 (spring 1993): 10–20. Quote is from Ruth Leigh, "Selling Globes and Atlases," reprinted from *Publishers' Weekly* (Chicago: Rand McNally & Co., 1929), in "History" Box, Rand McNally Collection.

19. Speech from 1913 reprinted in Thomas G. Paterson, ed., *Major Problems in American Foreign Policy: Volume I, to 1914* (Lexington, Mass.: D. C. Heath & Co., 1989), 505.

20. Wolfgang Scharfe, "German Atlas Development during the Nineteenth Century," in *Images of the World*, ed. Wolter and Grim, 207–32.

21. J. Paul Goode, *Goode's School Atlas* (Chicago: Rand McNally, 1932), xi–xii.

22. Initial editions of the atlas relied on Goode's homolographic projection, an interrupted Mollweide projection that is slightly different, though similar in appearance, to the homolosine projection.

23. Quote is from Bruce Grant, quoting W. G. North in interview with Andrew McNally III, in "History of Rand McNally and Company" (undated manuscript, c. 1956), 21.

24. See Alan K. Henrikson, "America's Changing Place in the World: From 'Periphery' to 'Centre'?" in *Centre and Periphery: Spatial Variation in Politics*, ed. Jean Gottmann (Beverly Hills, Calfi.: Sage Publications, 1980), 95 n. 9.

25. Exceptions occurred in National Geographic maps of 1935 and 1941, where the world was mapped as two separate circular hemispheres, necessarily placing the Western Hemisphere on the left side of the map. Generally the widest circulating atlases of the late nineteenth and early twentieth centuries—Century, Rand McNally, Hammond, Colton, and Mitchell, to name a few—placed the United States at the center of their Mercator-based world maps.

26. Goode, *Goode's School Atlas*, xiii.

27. "The Story of the Map," *National Geographic Magazine* 62, no. 12 (December 1932): 774.

28. Review of "Good's [*sic*] School Atlas" and "Hammond's Unabridged Atlas and Gazeteer of the World," *Historical Outlook* (May 1926): 249–50. All continents

were mapped on a scale of 1:40,000,000 and all regions as 1:16:000,000. Though comparatively more attention was paid to the United States and Europe, this still marks a relatively important shift when considered historically.

29. Grant, "History of Rand McNally," 20–21.

30. This is not to say that *Goode's* atlases were not successful: among high schools it has been the atlas of choice, and was extensively used in Army War Colleges during World War II.

31. Grant, quoting W. G. North in interview with Andrew McNally III, in "History of Rand McNally," 21.

32. See *Rand McNally World Atlas: Premier Edition* (1932, 1937). Rand McNally's other contemporary series—*World Atlas: Commonwealth Edition*—also did not include physical maps of any kind. C. S. Hammond delayed introducing physical maps into its bestselling atlases; as late as 1937, its *Modern Illustrated Atlas of the World* and *Unabridged Atlas and Gazetteer of the World* still included no physical maps.

33. Letter from Gilbert H. Grosvenor to Albert Holt Bumstead, 14 November 1918, in National Geographic Society, Records Division, GHG 11-10015.837.

34. For a more detailed explanation of the van der Grinten projection, see J. Paul Goode, "A New Method of Representing the Earth's Surface," *Journal of Geography* 4, no. 9 (November 1905): 369–73.

35. Jeremy Black, *Maps and Politics* (Chicago: University of Chicago Press, 1997), 31.

36. O. M. Miller, "Notes on Cylindrical World Map Projections," *Geographical Review* 32 (1942): 424; quoted in Frank Kuen Chung Wong, "World Map Projections in the United States from 1940–1960" (M.A. thesis, Syracuse University, 1965), 26, emphasis added.

37. Mark Twain, *Tom Sawyer Abroad* (New York: Harper and Brothers, 1896), 28–30.

38. Max Mayer, "Maps and Their Making," *Publishers' Weekly* 118 (1930): 976, 1663. Even J. Paul Goode recognized the excellence of the German and British atlases, yet he took care to argue for the democratizing influence of wax engraving on the American map industry.

39. "Brief on Behalf of Map Engravers and Publishers in Support of an Increase of Duties on Maps," Finance Committee, United States Senate, 71st Congress, "History" Box, Rand McNally Collection. H. B. Clow, "Why a Tariff on Maps? American Made Maps Superior to Foreign Made Maps," *Publishers' Weekly* 116 (1929): 2161–62.

40. John W. Hiltman (president of Appleton & Company), "As to the Imported Maps: The Harm Done to American Teaching by the Proposed Increase in Tariff," *Publishers' Weekly* 116 (1929): 1940, 2305.

41. Rand McNally & Company, "For Increasing Duties on Geographical Maps," Reply Brief of Rand McNally to D. Appleton & Company, Committee of Ways and Means, Tariff Readjustment 1929, 1–4; in "History" Box, Rand McNally Collection. Details of the Senate debate can be found in *Congressional Record: Proceedings and Debates of the Second Session of the Seventy-First Congress,* vol. 62, pt. 2 (Washington, D.C.: United States Government Printing Office, 1930), 1955–59.

42. Advertisement in *Review of Reviews*, November 1921. All Rand McNally advertisements collected in unmarked folder, Rand McNally Collection.

43. Advertisement in C. S. Hammond Company Records. Similarly, a C. S. Hammond advertisement fifteen years later for its unabridged atlas, *Our Planet*, boasted its inclusion of "a complete illustrated gazetteer, all of the world's famous landmarks, tables of all pertinent statistics, [and] an enormous index of over 50,000 place names." See *New York Times Book Review*, 24 February 1935, 23.

44. Advertisement in *The American*, September 1920. Other ads that referred to the accuracy of Rand McNally world atlases can be found in *World's Work*, November 1920; *Century*, March 1924; *Asia*, April 1924; *Atlantic Monthly*, August 1924; *Sunset*, March 1926; *Sunset*, November 1926; and *Review of Reviews*, May 1928.

45. *Red Book*, 1926. For other ads invoking the romance of the atlases, see *Sunset*, June 1926; *Sunset*, May 1927; *Review of Reviews*, September 1928; *World's Work*, December 1928. Advertisement with an interest in China was *World's Work*, October 1922.

46. Advertisement found in *Geographic*, February 1922.

47. Andrew McNally III, interview with the author, 14 June 1994, Chicago, Illinois.

48. Leigh, "Selling Globes and Atlases." See also "Who Will Sell Maps?" *Publishers' Weekly* 119 (1931): 2310; and "New Atlases in Demand," *Publishers' Weekly* 118 (1930): 2198.

49. Other institutional clients included The Pure Oil Company, Rock Island Lines, American Surety Company, Chicago North and Western Lines, the Bureau of Air Commerce, Pan American Airways, and the Monsanto Chemical Company. See Rand McNally Photograph Collections, illustrating different custom-made maps, globes, and displays for corporate clients, found in Box 7, Photographs and Scrapbooks, Rand McNally Collection.

Chapter Nine

1. "Roosevelt to Warn U.S. of Danger; Asks All to Trace Talk on Maps," *New York Times*, 21 February 1942, 1; "President's Speech Causes Great Demand for Maps," *Publishers' Weekly* 141, no. 9 (28 February 1942): 935–36. Roosevelt's own maps had come from the National Geographic Society's collection. See letter from Franklin Delano Roosevelt to the National Geographic Society, 24 December 1941, in Grosvenor Papers, Part I, Box 170, folder marked "Roosevelt, Franklin D. & Eleanor," Manuscript Division, Library of Congress.

2. Alan K. Henrikson, "The Map as an 'Idea': The Role of Cartographic Imagery during the Second World War," *American Cartographer* 2 (1975): 19–53. Quote is from James C. Malin, "Space and History: Reflections on the Closed-Space Doctrines of Turner and Mackinder and the Challenge of Those Ideas by the Air Age," *Agricultural History* 18 (April and July 1944): 107.

3. Vilhjalmur Stefansson, quoted in George B. Eisler's introduction to Erwin Raisz, *Atlas of Global Geography* (New York: Global Press, 1944), 5.

4. John McCannon, "To Storm the Arctic: Soviet Polar Exploration and Public Visions of Nature in the USSR, 1932–1939," *Ecumene* 2, no. 1 (1995): 22.

5. On sales see Memorandum from Harmon Woodworth to Andrew McNally III, undated [c. 1939], in Cartographic Publishing Box II, Rand McNally Collection, Newberry Library, Chicago [hereafter Rand McNally Collection]; "The Sudden Demand for Maps," *Publishers' Weekly* 136, no. 1 (9 September 1939): 881–83; and "The Boom in Maps," *Publishers' Weekly* 136, no. 2 (7 October 1939): 1423–24. This information was also broadcast on CBS radio, in an interview with Andrew McNally III by Bob Trout, 26 September 1939.

6. "Maps in Great Demand," *Publishers' Weekly* 141 (3 January 1942): 31–32. Richfield Gasoline and Richlube Motor Oils gave away to customers free maps of the European war theaters, with their own name emblazoned on the cover, and similar arrangements were made with other oil companies. See Rand McNally's *Mobilgas World Atlas* (1942), *Firestone World Atlas* (1942), *Richfield U.S. Defense Map* (1941), and *Sohio Standard Oil Training Camp Map of the U.S.* (1941), all uncatalogued in Rand McNally Collection. Rand McNally also sold their maps through magazines: *Life* reprinted the *Rand McNally Standard Map of Europe* in a two-page spread of the 25 September 1939 issue. The "Field Marshal's War Map" was advertised in the *New York Times*, *New Yorker*, *House Beautiful*, *National Geographic Magazine*, *Harpers*, and *House and Garden*, to name only a few.

7. Advertisement, circa early 1940s, in II Cartographic Publishing Box 10, Rand McNally Collection.

8. "Sudden Demand for Maps," 881; "Boom in Maps," 1424; "Maps," *Consumer Reports* 8, no. 11 (30 November 1943): 295.

9. See for example *Rand McNally World Atlas* (1942 edition), *Planters War Atlas* (1943), *Rand McNally Victory World Atlas* (1943 edition). Similar in style was *Cram's Global War Atlas* (1944). One of the few exceptions to this was the *Richfield European News Map* (Rand McNally, 1939) which included a detailed map of the relief, territorial occupation, and troop positions of the western front. All Rand McNally atlases in Rand McNally Collection.

10. The maps boosted sales substantially: in 1899 the Society needed only 1,417 copies of the Philippine map, while by 1950 it was printing 1.9 million copies of each map it created. In all, from 1899 to 1947 the Society created 101 color maps, and printed a total of 90 million copies. Gilbert H. Grosvenor, "Map Services of the National Geographic Society," in *The Round Earth on Flat Paper* (Washington, D.C.: National Geographic Society, 1947), 5, 8, 37.

11. The map was used in Army offices and on Naval ships, and was reprinted in *Yank*. See National Geographic Society, "Japan and Adjacent Regions of Asia and the Pacific Ocean." This projection, depicting distance and direction as true from the center of the map, was also used in both Society maps of Europe and the Near East, issued in May 1940 and in June 1943. Different projections were also used for the maps of Asia and the Soviet Union, issued in December 1942 and December 1944 respectively.

12. Erwin Raisz, "Outline of the History of American Cartography," *Isis* 26 (1937): 388. Albert Bumstead's innovative techniques were patented by the Society, and his legacy was carried on by his son as well as other notable cartographers at the Society. Grosvenor, "Map Services," 7–15.

13. *Newsweek*, 26 January 1942, 30.

14. "Japan and Adjacent Regions of Asia and the Pacific Ocean." Maps of the war theaters were also requested by the Air Force, as well as by the governments of Britain and South Africa. Grosvenor, "Map Services," 15, 20, 22. Perhaps the best known story about these maps was the Christmas exchange between Roosevelt and Churchill in 1943. Behind his desk, FDR kept a set of National Geographic maps during the war, and Churchill admired the collection so much that FDR requested that one be made for Churchill as a Christmas gift. Grosvenor, "Map Services," 20.

15. The letter is reprinted in its entirety in Grosvenor, "Map Services," 7, 14–17, 19–20. Statistics of the military's use of these maps also draw on the Society's own records.

16. The maps used in the Army and Navy educational corps were "The United States," "Germany and Its Approaches," and the eastern portion of the Society's "Asia" map. The maps reproduced in Yank were "Central Europe and the Mediterranean," "The Pacific Ocean and Bay of Bengal," "Germany and Its Approaches," and "Japan and Adjacent Regions of Asia and the Pacific Ocean." See Yank, 18 February 1944; 25 February 1944; 17 March 1944; 12 January 1945. Other maps, also intended as pull-outs, were borrowed from Newsmap, a set of maps produced during the war for research, instruction, and public information. The Yank maps thoroughly integrated the innovations of journalistic cartography, including polar projections and aerial perspectives that highlighted the spherical nature of the earth and the global nature of the war. Many of these maps were signed by Sargeant F. Brandt. Those reprinted from Newsmap included one of France and the Low Countries, in the 30 June 1944 issue, and of Germany and its borders, in the 14 September 1944 issue.

17. New York Times, 1942, quoted in Grosvenor, "Map Services," 31. The extent of the journalistic community's admiration for the Society's maps was demonstrated on April 8, 1944, when the Associated Press, the United Press, and International News Service recognized the Society's place-name spellings as official. The debated need for a "standard nomenclature" had ebbed and flowed over the previous five decades, though standardization remained elusive.

18. From 1860 to 1930 Monmonier found that the New York Times on average ran less than one article with a map in each issue. By 1940, that average had risen to three per issue, with episodic bursts around the Spanish-American War and World War I, and a qualitative rise in the 1940s due to the war and aviation. In 1940, 39 percent of Times maps were related to military conflicts and issues of defense and geopolitics; by 1950 that percentage had peaked at 45 percent. Mark Monmonier, "Maps in The New York Times, 1860–1980: A Study in the History of Journalistic Cartography," Proceedings of the Pennsylvania Academy of Science 58 (1984): 79–83; and Monmonier, "Maps in The Times (of London) and The New York Times, 1870–1980: A Cross-National Study in Journalistic Cartography," Proceedings of the Pennsylvania Academy of Science 59 (1985): 61–66.

19. Notable among these journalistic cartographers were Richard Edes Harrison, Joseph Johnson, and LeRoy Appleton at the Fortune, Emil Herlin at The New York Times, Van Swearingen at The Chicago Tribune, F. E. Manning at The Chicago Sun, Robert Chapin at Time, and especially Antonio Petrucelli, H. C. Detje, and Erwin Raisz, whose popular maps were reprinted in many newspapers and periodicals. Raisz was unique for his professional cartographic training; most were graphic artists. Walter W. Ristow, "Journalistic Cartography," Surveying and Mapping 27 (October–

December 1957): 369; and Mark Monmonier, *Maps with the News: The Development of American Journalistic Cartography* (Chicago: University of Chicago Press, 1989), 18.

20. For a more detailed discussion of Harrison's work, see Susan Schulten, "Richard Edes Harrison and the Challenge to American Cartography," *Imago Mundi* 50 (1998): 174–88.

21. Richard Edes Harrison to editors of *Fortune*, 20 October 1942, in Richard Edes Harrison Collection, Geography and Map Division, Library of Congress. Much of this discussion also relies on my interview with Harrison, New York City, 14 October 1993.

22. Richard Edes Harrison, uncatalogued notes of 3 August 1944, in Harrison Collection.

23. Richard Edes Harrison, "Making Maps Tell the Truth," *Travel* 80 (1942): 10–13. *The World Divided* first appeared in *Fortune* (August 1941): 48–49. These maps were also printed in "Perspective Maps: Harrison Atlas Gives Fresh New Look to Old World," *Life* (28 February 1944): 56–61.

24. Richard Edes Harrison, *Look at the World: The FORTUNE Atlas for World Strategy* (New York: Knopf, 1944).

25. Alexander De Seversky, *Victory through Air Power* (New York: Simon and Schuster, 1942), 7.

26. Richard Edes Harrison, notes, in "Clients" folder, Harrison Collection, Geography and Map Division, Library of Congress.

27. Wellman Chamberlin, "The Round Earth on Flat Paper: A Description of Map Projections Used by Cartographers," in *The Round Earth on Flat Paper*, 52–55.

28. Letter from Charles Colby to Richard Edes Harrison, 1 October 1941, in Harrison Collection.

29. Letter from Richard Edes Harrison to George B. Cressey, Department of Geology and Geography, Syracuse University, 8 December 1941, in Harrison Collection.

30. *First Six Months of War* (New York: Associated Press, 1940); *Newsweek Global War: Four Years of World Conflict, 1939–1943* (Newsweek, 1944); *Time's Atlas of the WAR* (Time, c. 1943). Another impressive series was the "Esso War Maps," which emphasized the principles of cartographic projection and scale for the general public.

31. The American Geographical Society was also an important source of maps during the war, but because its audience was limited to geographers and government officials it is not included in this study. For a history of the American Geographical Society, see John K. Wright, *Geography in the Making: The History of the American Geographical Society* (New York: American Geographical Society, 1957).

32. Helmuth Bay, "War News via Maps," *Wilson Library Bulletin* 18 (November 1943): 221.

33. FDR fireside chat of 2 February 1942; *New York Times*, 21 February 1942, 1.

34. Helmuth Bay, "The History and Technique of Mapmaking," in *Eighth of the R. R. Bowker Memorial Lectures* (New York: New York Public Library, 1943); Bay, "War News via Maps."

35. "Maps: Global War Teaches Global Cartography," *Life*, 3 August 1942, 57.

Letter from Harrison to his editors at *Fortune*, 20 October 1942, Harrison Collection. The wartime literature on map projections is voluminous. For examples, see W. J. Luyten, "Those Misleading New Maps," *Harper's Magazine* 187 (October 1943): 447–49; "War on the Visual Front," *American Scholar* (October 1942); Hans Weigert, "Maps Are Weapons," *Survey Graphic* 30 (October 1941): 528–30; CBS, "Americans at Work," radio program no. 34, "The Map Maker," Columbia Broadcasting System, written by Margaret Lewerth and directed by Brewster Morgan, produced by CBS Adult Education Board; II Cartographic Publishing Box 1, Rand McNally Collection. WOR radio program, Bessie Beattie, aired 26 June 1944, reviewing Harrison's *Look at the World* atlas. "City's Schools Will Use Flat 'Globe' Maps in Fall," *New York Times*, 2 August 1943, 12. *New York Times*, 23 October 1943, 14. On map projections see Irving Fisher and O. M. Miller, *World Maps and Globes* (New York: Essential Books, 1944); David Greenhood, *Down to Earth* (New York: Holiday House, 1944).

36. Those recommended included the *Polar View Globe*, *Rand McNally Air-Age Globe*, *Air Chief Polar View Globe*, and *Airways Globe*. The company introduced the cradle mounting for the twelve-inch globe in 1939. All recommendations from "Maps," *Consumer Reports*, 295. Significantly, the demand for globes during the war never approximated that for atlases and maps because globes allowed minimal detail, and virtually no means to follow the war. Ironically, though the trend of the period was to call for a new "global view," during the war itself globes were particularly scarce because many globe plants were converted to manufacture of helmets and safety visors for use by armed forces and war workers, and manufacturers lost both employees and materials to the war effort. "The Sudden Demand for Maps," 83; "Maps," 241–42; "Globes Chronology," unauthored record of Rand McNally globe production, in II Cartographic Publishing Box 1, Rand McNally Collection.

37. During the 1940s, Rand McNally advertised in magazines such as *Business Week*, *Time*, and *The New Yorker*. Ads found in *Time National Advertising Anthology*, in "Cartographic Advertisements" Box, Rand McNally Collection.

38. Richard Edes Harrison and Hans W. Weigert, "World View and Strategy," in *Compass of the World: A Symposium on Political Geography*, ed. Hans W. Weigert and Vihjalmur Stefansson (New York: Macmillan, 1944), 74–75; Gearóid Ó Tuathail, "It's Smart to Be Geopolitical," in *Critical Geopolitics: The Politics of Writing Global Space* (Minneapolis: University of Minnesota Press, 1996).

39. In an interview with the editor of the atlas, Duncan Fitchet, the *New Yorker* lauded the company's ability to respond to changing conditions wrought by the war. "From Scratch," *New Yorker* 25 (29 October 1949): 19–20. See also Harland Manchester, "The World of Rand McNally," *Saturday Review of Literature* 33 (21 January 1950): 35–37. *Publishers' Weekly* also praised the atlas in "Rand McNally Wins Carey-Thomas Award for World Atlas," *Publishers' Weekly* 157 (4 February 1950): 783–87.

40. "The Rand McNally *Cosmopolitan World Atlas*: How and Why It Was Made," Cartographic Publishing Atlases, Box 4, Rand McNally Collection.

41. Hammond spent nearly as much as Rand McNally on cartographic revisions after the war, reporting a cost of $3,000 for each of the 104 new maps found in the 1950 *Library World Atlas*. The company also decided to switch to offset printing; when their plate finisher died in 1938 he took with him the historically informal and secretive process of wax engraving, ending the nearly seventy-five-year reign of the tech-

nique in American cartography. Most of the changes in the *Cosmopolitan Atlas* were adopted by Hammond's postwar atlases as well. Firms founded in the twentieth century, such as Denoyer-Geppert, General Drafting, A. J. Nystrom, and the American Map Company, had not invested so heavily in the wax process, and had been using these newer printing techniques from the outset. Woodward, *All-American Map,* 38–41.

42. "Old independent nations" were identified as those charter members of the United Nations; "young independent nations" were World War I mandates or colonies recently liberated; "unstable dependent areas" included former colonies demanding independence or self-government; "stable dependent areas" referred to those areas with "varying degrees of political unrest, but not actively in revolt"; "dependent areas under United Nations Trusteeships" included pre–World War I German colonies in Africa and the Pacific as well as the former Japanese mandated islands; "areas under military occupation" indicated Germany, Austria, the Italian colonies, Japan, and Korea; and "recently transferred territories" included those acquired during or after the war by surrender agreements, treaties, or unilateral action. See *Cosmopolitan Atlas of the World,* xiv.

43. The new map of the Atlantic was drawn on a similar projection, pulling Europe closer to the United States. This new map corresponded to what Henrikson has identified as a general conceptual change among foreign policy analysts toward understanding the Atlantic as a body of water surrounded by land, a "maritime sphere defined by the several coastlines of the North Atlantic, an 'Atlantic Area' cradling an infant 'Atlantic Community.'" Henrikson, "Map as an 'Idea,'" 31–33.

44. Important changes also appeared in regional mapping. The map of Europe was widened to include western Asia, even though this diminished the relative size of western Europe. Actual boundary changes were most important along the Soviet Union's western border, between Germany and Poland, Italy and Yugoslavia, and Bulgaria and Rumania. Compare these minor adjustments to the massive changes after World War I, when Europe created Czechoslovakia, Poland, Finland, Yugoslavia, Albania, three Baltic States, and an independent Austria and Hungary. Outside of Europe, Rand McNally took account of new republics—Burma, the Philippines, Iceland, Lebanon Syria, and Mongolia, as well as the kingdoms of Trans-Jordan and Yemen. Though the United States recognized Israel, and the atlas identified its capital as Tel Aviv, its borders were not drawn on the map. See "From Scratch," cited in note 39 above.

45. "Rand McNally Wins Carey-Thomas Award," 783.

46. For the explanation of scale, see *Cosmopolitan Atlas,* xii.

47. Percentages do not total 100 because of polar maps and world maps.

Epilogue

1. Neil Smith, *The Geographic Pivot of History: Isaiah Bowman and the Geography of the American Century* (Baltimore: Johns Hopkins University Press, forthcoming), 25, emphasis added.

2. Benedict Anderson, *Imagined Communities* (London: Verso, 1991), p. 173.

SELECT BIBLIOGRAPHY

Manuscript Collections

Alexander Graham Bell Papers, Manuscript Division, Library of Congress, Washington, D.C.

Gilbert Grosvenor Papers, Manuscript Division, Library of Congress, Washington, D.C.

C. S. Hammond Company Records, C. S. Hammond, Maplewood, New Jersey.

Richard Edes Harrison Collection, Geography and Map Division, Library of Congress, Washington, D.C.

W J McGee Papers, Manuscript Division, Library of Congress, Washington, D.C.

Rand McNally Collection, Newberry Library, Chicago.

National Geographic Society Archives, Records Division, National Geographic Society, Washington, D.C.

Interviews with the Author

Stephen Gale, Philadelphia, December 12, 1993.

Caleb D. Hammond, Maplewood, New Jersey, March 3, 1995.

Chauncy Harris, Philadelphia, April 4, 1994.

Richard Edes Harrison, New York, October 14, 1993.

Andrew McNally III, Chicago, June 10, 1994.

Atlases and Maps

Rand McNally & Company

1885, 1898	*New Household Atlas of the World*
1887–1939	*Pocket Atlas of the World*
1888, 1899	*New Family Atlas of the World*
1894	*Library Atlas of the World*
1895–1905	*New General Atlas of the World*
1896–1900	*Pictorial Atlas of the World*
1897	*World's Peoples & Countries*
1898	*Rand McNally War Atlas*
1898	*History of the Spanish-American War with Handy Atlas Maps*

1898, 1925	Columbian Atlas of the World
1898–1899	Encyclopedia & Atlas of the World
1898–1918	New Imperial Atlas of the World
1899	New Concise Atlas of the World
1899	Atlas of Two Wars . . . Philippine Islands and South Africa
1900	Wms & Peters' Atlas of China . . . Pertaining to the Present Crisis
1904	Russo-Japanese War Atlas
1905–1913	World & Its Peoples
1913	Atlas of the Mexican Revolution
1914	Atlas of the European Conflict
1914, 1916	New Family Atlas of the World
1915	Graphic Representation of the Battle Fields of Today
1915, 1921–22	Ideal Atlas of the World
1917	Rand McNally Atlas of the World War
1918	Atlas of the World War
1917–1918	New Reference Atlas of the World and the War
c. 1920	Battle Fields of Yesterday
1921	Atlas of Reconstruction for Schools
c. 1922	World's Resources and Where to Find Them
1920	Foreign Trade Atlas: Three Americas
1922–1992	Goode's School Atlas
1923–1978	Premier Atlas of the World
1927–1947	World Atlas: International Edition
1942–1958	Popular World Atlas
1939, 1942	Rand McNally World Atlas, Regular & European War Edition
1942, 1943	Rand McNally Victory World Atlas
1943	Planters Presents Our Fighting Forces
1949–1978	Cosmopolitan Atlas of the World
1949	Fifty Fabulous Years, 1900–1950

George F. Cram & Company

1883, 1885	Cram's Unrivaled Family Atlas
1886	The People's Illustrated and Descriptive Family Atlas
1887–1952	Cram's Unrivaled Atlas of the World
1891–1910	Cram's Railway Atlas
1891–1952	Cram's Unrivaled Family Atlas of the World
1897	New Official Map of Alaska and the Klondike Gold Fields
1899	Cram's Library and School Map of the World
1902	Cram's Ideal Reference Atlas of the World
1915	Cram's Atlas of the War in Europe
1922	Cram's Modern Reference Atlas of the World
1942	Cram's Global War Atlas

C. S. Hammond & Company

1905–1937	*Hammond's Modern Atlas of the World*
1907–1913	*Hammond's Handy Atlas of the World*
1924	*Hammond's Unabridged Atlas and Gazetteer of the World*
1937–1938	*Hammond's Modern Illustrated Atlas of the World*
1947	*Hammond's New World Atlas*
1950	*Hammond's Library World Atlas*

Richard Edes Harrison Maps in *Fortune* Magazine

November 1935	*Vulture's View*
April 1936	*World Airways*
September 1936	*The Expanding Empire of Japan*
February 1937	*The Great Seaports*
October 1939	*Europe 1939*
July 1940	*The Great Lakes*
September 1940	*Atlas for the U.S. Citizen*
April 1941	*China as Seen from Guam*
July 1941	*Russia*
August 1941	*The World Divided*
November 1941	*Africa*
March 1942	*One World—One War*
June 1942	*Atlantic Arena*
July 1942	*Arctic Arena*
September 1942	*Pacific Arena*
November 1942	*The Middle East*
December 1942	*Japan's Empire*
January 1943	*The Not-so-Soft Underside (Europe Seen from Africa)*
May 1943	*Great Circle Airways*
December 1943	*Approaches to Tokyo*
February 1944	*The Caribbean*
April 1944	*Japan*
October 1944	*Colonial Empire—Africa*
October 1944	*Proposed Regional Groupings in South East Asia*
November 1944	*Philippines*
January 1945	*Economic Russia*
October 1945	*Sugar: Where It Grows and Where It Goes*

National Geographic Society Maps (Selected)

May 1918	*Map of the Western Theatre of War*
December 1932	*The World*
October 1939	*Central Europe and the Mediterranean*
May 1940	*Europe and the Near East*
March 1941	*Indian Ocean*

September 1941	*Atlantic Ocean*
December 1942	*Asia and Adjacent Areas*
June 1943	*Europe and the Near East*
September 1943	*Pacific Ocean and the Bay of Bengal*
April 1944	*Japan and Adjacent Regions of Asia and the Pacific Ocean*
July 1944	*Germany and Its Approaches*
October 1944	*South East Asia and the Pacific Islands*
December 1944	*Union of Soviet Socialist Republics*
March 1945	*The Philippines*

Other Atlases

1895–1922	*Steiler's Hand-Atlas* (J. Perthes)
1912–1947	*Citizen's Atlas* (Bartholomew)
1931	*Philips' International Atlas* (London: George Philip & Son)
1934	*Philips' Record Atlas* (London Geographical Institute)
1940	*First Six Months of War* (Associated Press)
c. 1943	*Time's Atlas of the* WAR (Time)
1944	*Newsweek Global War: Four Years of World Conflict*
1944	*A War Atlas for Americans* (Simon and Schuster)
1944	*Erwin Raisz's Atlas of Global Geography* (Global Press Corp.)

Books and Articles

A Business Tour of Chicago Depicting Fifty Years' Progress. Sights and Scenes in the Great City, Her Growing Industries and Commercial Development, Historical and Descriptive. Prominent Places and People . . . for Popular Distribution. Chicago: E. E. Barton, 1887.

"A Loss to Publishing." *Publishers' Weekly* 116 (1929): 2926.

Aay, Henry. "Conceptual Change and the Growth of Geographic Knowledge." Ph.D. diss., Clark University, 1978.

Abrams, Paul P. "Academic Geography in America: An Overview." *Reviews in American History* 4 (1975): 46–52.

Abramson, Howard S. *National Geographic: Behind America's Lens on the World*. New York: Crown Publishers, 1987.

Ackerman, Edward A. "Geographic Training, Wartime Research, and Immediate Professional Objectives." *Annals of the Association of American Geographers* 35 (1945): 121–43.

Adams, Cyrus C. "Maps and Mapmaking." *Bulletin of the American Geographical Society* 44 (1912): 198–201.

Ager, John. "Maps & Propaganda." *Society of University Cartographers Bulletin* 11 (1977): 1–14.

Akerman, James. "On the Shoulders of a Titan: Viewing the World of the Past in Atlas Structure." Ph.D. diss., Pennsylvania State University, 1991.

―――. "Blazing a Well Worn Path: Cartographic Commercialism, Highway Promotion, and Auto Tourism in the United States, 1880–1930." *Cartographica* 30, no. 1 (spring 1993): 10–20.

―――. "Selling Maps, Selling Highways: Rand McNally's 'Blazed Trails' Program." *Imago Mundi* 45 (1993): 77–89.

Aldridge, Dorothy. "Present Status and Current Trends of Geographic Instruction in Massachusetts High Schools." M.A. thesis, Clark University, 1948.

"The American Geographical Society's Contribution to the Peace Conference." *Geographical Review* 7 (1919): 1–10.

Anderson, Benedict. *Imagined Communities: Reflections on the Origin and Spread of Nationalism.* London: Verso, 1983.

Anderson, Howard R. "Offerings and Registrations in Social Studies." *Social Education* 14 (1950): 73–75.

Andrews, Jane. *Geographical Plays.* Boston: Ginn and Company, 1880, 1894.

Annals of the Association of American Geographers 37 (1947): 1–15.

Argenbright, Robert. "Bowman's New World: World Power and Political Geography." Masters thesis, University of California, Berkeley, 1985.

Armstrong, T. H. "Relation of Geography to the Other Studies in the Elementary Course of Study." *Education* 23 (1903): 331–36.

Army Map Service. "Arms and the Map: Military Mapping by the Army Map Service." *Print* 4, no. 2 (spring 1946): 3–16.

Arnold, Edmund R. "The American Geographical Society Library, Map and Photograph Collection: A History, 1951–1978." Ed.D. diss., University of Pittsburgh, 1985.

Atwood, Wallace W. "The New Meaning of Geography in American Education." *School and Society* 13 (1921): 211–18.

―――. *The World At Work.* Boston: Ginn and Company, 1931.

―――. *Graduate School of Geography in Clark University, 1920–1945.* Worcester: Clark University, 1946.

―――. "New Meaning of Geography in World Education." *Journal of Geography* 46 (January 1947).

Atwood, Wallace, and Helen G. Thomas. *The Earth and Its Peoples: Nations Beyond the Seas.* Boston: Ginn and Company, 1930.

Bagley, William C. "Functions of Geography in the Elementary School: A Study in Educational Values." *Journal of Geography* 3 (1904): 222–33.

Baker, J.N.L. *The History of Geography.* New York: Barnes and Noble, 1963.

Balliet, Thomas M. "Notes on Teaching Geography." *New York Teachers' Monographs* 2, no. 1 (1899): 48–51.

Barrows, Harlan H. "A Textbook of Principles of Human Geography." *Geographical Review* 12 (January 1922): 157–60.

―――. "Geography as Human Ecology." *Annals of the Association of American Geographers* 13 (1923): 1–14.

———. "The Purpose of Geography Teaching." *Journal of Geography* 20 (1921): 151–54.

Barrows, Harlan H., and Edith Putnam Parker. *Countries Throughout the World*. New York: Silver Burdett, 1933, 1938, 1941.

Barton, Ersella M., and Thomas F. Barton. "High School Geography in Southern Illinois." *Transactions of the Illinois State Academy of Science* 33 (December 1940): 131–33.

Bassett, Thomas J. "Cartography and Empire Building in Nineteenth-Century West Africa." *Geographical Review* 84, no. 3 (July 1994): 316–35.

Bassin, Mark. "Imperialism and the Nation State in Friedrich Ratzel's Political Geography." *Progress in Human Geography* 11 (1987): 123–32.

———. "Race Contra Space: The Conflict between German Geopolitik and National Socialism." *Political Geography Quarterly* 6, no. 2 (1987): 115–34.

Bay, Helmuth. "The History and Technique of Map Making." *Eighth of the R. R. Bowker Memorial Lectures*. New York: New York Public Library, 1943.

———. "War News Via Maps." *Wilson Library Bulletin* 18 (November 1943): 221–25.

Beck, Joanna E. "Environmental Determinism in Twentieth–Century American Geography: Reflections in the Professional Journals." Ph.D. diss., University of California, Berkeley, 1985.

Becker, Christian. "History of the Development of the Course of Study of Geography in New York City High Schools, 1898–1953." Ed.D. diss., New York University, 1954.

Black, Jeremy. *Maps and Politics*. Chicago: University of Chicago Press, 1997.

Blodgett, James H. "School Text-Books in Geography." *The Journal of School Geography* 3 (April 1899): 138–45.

Blouet, Brian W. "The Political Career of Sir Halford Mackinder." *Political Geography Quarterly* 6 (1987): 355–67.

———. *Halford Mackinder, A Biography*. College Station: Texas A&M University Press, 1987.

Blouet, Brian, ed. *Origins of Academic Geography in the United States*. Hamden, Conn.: Archon Books, 1981.

"Bookstores All Over Country Feature Maps and Globes." *Publishers' Weekly* 145, no. 2 (17 June 1944): 2239–43.

"The Boom in Maps." *Publishers' Weekly* 136, no. 2 (7 October 1939): 1423–24.

Bosse, David. *Civil War Newspaper Maps: A Cartobibliography of the Northern Daily Press*. Westport, Conn.: Greenwood Press, 1993.

Bourne, Henry Eldridge. *The Teaching of History and Civics in the Elementary and the Secondary Schools*. New York: Longmans Green & Company, 1900.

Bourne, Randolph S. *The Gary Schools*. Cambridge: MIT Press, 1970 [1916].

Bowman, Isaiah. *The New World: Problems in Political Geography*. Yonkers-on-Hudson, N.Y.: World Book Company, 1921.

———. *Geography in Relation to the Social Sciences*, Part V of the Report of the Commission on the Social Studies of the American Historical Association. New York: Charles Scribner's Sons, 1934.

———. "Geography vs. Geopolitics." *Geographical Review* 32 (1942): 646–58.

———. "Political Geography of Power." *Geographical Review* 32 (1942): 349–52.

Bradley, John H. *World Geography*. Boston: Ginn and Company, 1945.

Breasted, Charles. *Pioneer to the Past: The Story of James Henry Breasted, Archaeologist*. Chicago: University of Chicago Press, 1977 [1943].

Breitbart, Eric. *A World on Display: Photographs from the St. Louis World's Fair*. Albuquerque: University of New Mexico Press, 1997.

Brigham, Albert Perry. *Geographic Influences in American History*. Boston: Ginn and Company, 1903.

———. *Commercial Geography*. Boston: Ginn and Company, 1911.

———. "Geography After the War." *Educational Review* 57 (1919): 277–85.

———. "Geography and the War." *Journal of Geography* 19 (1920): 89–102.

———. "The Association of American Geographers, 1903–1923." *Annals of the Association of American Geographers* 14 (1924): 109–16.

Brigham, Albert Perry, and Richard Elwood Dodge. "Nineteenth Century Textbooks of Geography." In *The Teaching of Geography*, by the National Society for the Study of Education. Bloomington, Ill.: NSSE.

Brigham, Albert Perry, and Charles T. McFarlane. *Our World and Ourselves: Our Neighbors Near and Far*. Boston: Ginn and Company, 1933.

Bronson, Judith Conoyer. "Ellen Semple: Contributions to the History of American Geography." Ph.D. diss., St. Louis University, 1973.

Brückner, Martin. "Lessons in Geography: Maps, Spellers, and Other Grammars of Nationalism in the Early Republic." *American Quarterly* 51, no. 2 (June 1999): 311–43.

Burdick, Alger Ernest. "The Contributions of Albert Perry Brigham to Geographic Education." Ed.D. diss., George Peabody College for Teachers, June 1951.

Buxbaum, Edwin C. *Collector's Guide to the National Geographic*. Wilmington, Del.: Buxbaum, 1962.

Buxton, Frank. *The Big Broadcast, 1920–1950*. New York: Viking, 1972 [1966].

Cadugan, William. "An Analysis of Economic Geography Texts, 1891–1956." Ed.D. diss., University of Pittsburgh, 1958.

Calhoun, Daniel. "Eyes for the Jacksonian World: William C. Woodbridge and Emma Willard." *Journal of the Early Republic* 4 (spring 1984): 1–26.

Cameron, Ian. *To the Farthest Ends of the Earth: The History of the Royal Geographical Society, 1830–1980*. London: MacDonald, 1980.

Capel, Horatio. "Institutionalization of Geography and Strategies of Change." In *Geography, Ideology, and Social Concern*, edited by David Stoddart. Totowa, N.J.: Barnes and Noble, 1981.

Cappon, Lester J. "The Historical Map in American Atlases." *Annals of the Association of American Geographers* 69 (1979): 622–34.

Carey, Everett P. "General Science in Relation to Physical Geography." *Journal of Geography* 10 (1911): 62–66.

Carpenter, Charles. *History of American Schoolbooks*. Philadelphia: University of Pennsylvania Press, 1963.

Carpenter, Frank G. *Asia: A Geographical Reader*. New York: American Book Company, 1897.

———. *Europe: A Geographical Reader*. New York: American Book Company, 1902.

———. *Australia, Our Colonies, and other Islands of the Seas*. New York: American Book Company, 1904.

Chamberlain, James F. "Geography in the Life of the Pupil." *Proceedings of the NEA* (1907): 497–502.

———. "Report of the Committee on Secondary School Geography." In *Addresses and Proceedings of the NEA*. Chicago: University of Chicago Press, 1909.

———. *Air-Age Geography and Society*. Chicago: J. B. Lippincott Company, 1945.

Chamberlin, Wellman. *The Round Earth on Flat Paper: A Description of Map Projections Used by Cartographers*. Washington, D.C.: National Geographic Society, 1950 [1947].

Chorley, R. J., with R. P. Beckinsale and A. J. Dunn. *The History of the Study of Landforms, or the Development of Geomorphology. Volume 2: The Life and Work of William Morris Davis*. London: Methuen, 1973.

Clark, Rose B. "Geography in the Schools of Europe." In *Geography in Relation to the Social Sciences*, by Isaiah Bowman. New York: Charles Scribner's Sons, 1934.

Cleveland, Reginald, and Leslie E. Neville. *The Coming Air Age*. New York: McGraw-Hill, 1944.

Clow, H. B. "Why a Tariff on Maps? American Made Maps Superior to Foreign Made Maps." *Publishers' Weekly* 116 (1929): 2161–62.

Cohen, Patricia Cline. "Statistics and the State: Changing Social Thought and the Emergence of a Quantitative Mentality in America, 1790 to 1820," *William and Mary Quarterly* 38, no. 1 (1981): 35–55.

Colby, Charles. "Changing Currents of Geographic Thought in America." *Annals of the Association of American Geographers* 26 (1936): 1–37.

Colby, Charles, and Alice Foster. *Economic Geography*. Boston: Ginn and Company, 1940.

Collier, James E. "Geography in the High Schools of Arkansas." *Journal of Geography* 42 (1943): 134–44.

Collins, Jane, and Catherine Lutz. "Becoming America's Lens on the World: *National Geographic* in the Twentieth Century." *South Atlantic Quarterly* 91, no. 1 (winter 1992): 161–91.

Committee on Training and Standards in the Geographic Profession. "Lessons from the War-Time Experience for Improving Graduate Training for Geographic Research." *Annals of the Association of American Geographers* 36 (1946): 195–214.

Congdon, Lenore. *Schutze's Amusing Geography and System of Mapmaking*. San Francisco: Whitaker and Ray, 1899.

Conzen, Michael, ed. *Chicago Mapmakers: Essays on the Rise of the City's Map Trade*. Chicago: Chicago Historical Society, 1984.

Corn, Joseph J. *Winged Gospel: America's Romance with Aviation, 1900–1950*. New York: Oxford University Press, 1983.

Cremin, Lawrence. *A History of Education in American Culture*. New York: Holt, 1953.

———. *The Transformation of the Schools*. New York: Knopf, 1961.

Cubberley, Ellwood P. *The Portland Survey: A Textbook on City School Administration Based on a Concrete Study*. Yonkers-on-Hudson, N.Y.: World Book Company, 1916.

Culler, Ned. "Development of American Geography Textbooks from 1840–1890." Ph.D. diss., University of Pittsburgh, 1945.

Cutshall, Alden. "High School Geography in Illinois." *School Science and Mathematics* (March 1942): 560–64.

Danzer, Gerald A. "George F. Cram and the American Perception of Space." In *Chicago Mapmakers: Essays on the Rise of the City's Map Trade*, edited by Michael Conzen. Chicago: Chicago Historical Society, 1984.

Darby, H. C. "Academic Geography in Britain: 1918–1946." *Transactions of the Institute of British Geographers* 4 (1947): 187–94.

David, Tudor. "Against Geography." *Universities Quarterly* 12 (1957–58): 261–73.

Davis, Mary, and Charles Deane. *Elementary Inductive Geography*. New York: Potter and Putnam, 1900.

Davis, William Morris. "Teaching of Geography." *Educational Review* 3 (1892): 417–26.

———. *Physical Geography*. Boston: Ginn and Company, 1898.

———. "The Present Trend of Geography." *111th Annual Report of the University of the State of New York* (1898): 196.

———. "Physical Geography in the High School." *School Review* 8 (1900): 388–404, 449–56.

———. "Systematic Geography." *Proceedings of the Philosophical Society* 40 (3 April 1902): 253–58.

———. "Geography in the United States." *Science* 19 (1904): 120–32, 178–86.

———. "Geography in the United States." *Scientific American Supplement* no. 1463 (16 January 1904): 23450–51.

———. "The Opportunity for the Association of American Geographers." *Bulletin of the American Geographical Society* 37 (1905): 84–86.

———. "The Progress of Geography in the United States." *Annals of the Association of American Geographers* 14 (1924): 159–215.

De Seversky, Alexander P. *Victory through Air Power*. New York: Simon and Schuster, 1942.

Deasy, George. "Training, Professional Work and Military Experience of Geographers, 1942–1947." *The Professional Geographer* 6 (December 1947): 1–14.

DeBres, Karen. "George Renner and the Great Map Scandal." *Political Geography Quarterly* 5, no. 4 (October 1986): 385–94.

———. "An Early Frost: Geography in Teachers College, Columbia and Columbia University, 1896–1942." *Geographical Journal* 155, no. 3 (1989): 392–402.

Deetz, Charles H., and Oscar S. Adams. *Elements of Map Projection*. Coast and Geodetic Survey, Special Publication no. 68. Washington, D.C.: Department of Commerce, 1934.

DeLima, Agnes. *Our Enemy the Child*. New York: New Republic, Inc., 1926. Reprint, New York: Arno Press, 1969.

Dickinson, Robert E. *Makers of Modern Geography*. London: Routledge and Kegan Paul, 1969.

———. *Regional Concept: Anglo American Leaders*. London: Routledge and Kegan Paul, 1969.

Dodge, Richard Elwood. *Reader in Physical Geography for Beginners*. New York: Longmans Green and Company, 1900.

———. *Elementary Geography: Part Two—Comparative Geography of the Continents*. Chicago: Rand McNally, 1904, 1907, 1916.

———. "Report of the Committee on Geography for Secondary Schools." *Journal of Geography* 8 (1910): 159–65.

Dodge, Richard Elwood, and William E. Grady. *World Relations and the Continents*. Chicago: Rand McNally, 1914, 1922.

Dodge, Richard Elwood, and Clara Kirchwey. *The Teaching of Geography in Elementary Schools*. Chicago: Rand McNally, 1913.

———. "A Study of Geography in Normal Schools." *Teachers College Review* 15, no. 2 (1914): 71–82.

Dodge, Richard Elwood, and Earl Emmet Lackey. *Dodge-Lackey Elementary Geography*. Chicago: Rand McNally, 1927.

———. *Dodge-Lackey Advanced Geography*. Chicago: Rand McNally, 1928.

Donoghue, John H. "Maps Must be Made by the Millions." *Military Engineer* 34 (1942): 427–30.

Dorflinger, Johannes. "Austrian Atlases of the Late Nineteenth and Early Twentieth Centuries." In *Images of the World: The Atlas through History*, edited by John A. Wolter and Ronald E. Grim. New York: McGraw-Hill, 1997.

Driver, Felix. "Geography's Empire: Histories of Geographical Knowledge." *Environment and Planning D: Society and Space* 10 (1992): 23–40.

Dryer, Charles Redway. *High School Geography*. New York: American Book Company, 1912.

———. "A Century of Geographical Education." *Annals of the Association of American Geographers* 14 (1924): 117–49.

Dunbar, Gary S. "The Rival Geographical Societies of *Fin-de-Siècle* San Francisco." *Yearbook of the Association of Pacific Coast Geographers* 40 (1978): 57–63.

———. "Geography in the Bellwether Universities of the United States." *AREA* (Institute of British Geographers) 18, no. 1 (1985): 25–33.

Dunbar, Gary S., ed. *Modern Geography: An Encyclopedic Survey*. New York: Garland Publishers, 1991.

Dunning, John. *Tune in Yesterday: The Ultimate Encyclopedia of Old-Time Radio, 1925–1976*. Englewood Cliffs, N.J.: Prentice Hall, 1976.

Dupree, A. Hunter. *Science in the Federal Government: A History of Policies and Activities*. Baltimore: Johns Hopkins University Press, 1986.

Earle, Carville. *Geographical Inquiry and American Historical Problems*. Palo Alto, Calif.: Stanford University Press, 1992.

Edney, Matthew H. "Politics, Science, and Government Mapping Policy in the United States, 1800–1925." *American Cartographer* 13, no. 4 (1986): 295–306.

———. "Cartographic Culture and Nationalism in the Early United States: Benjamin Vaughan and the Choice for a Prime Meridian, 1811." *Journal of Historical Geography* 20, no. 4 (1994): 384–95.

———. *Mapping an Empire: The Geographical Construction of British India, 1765–1843*. Chicago: University of Chicago Press, 1997.

Elson, Ruth Miller. *Guardians of Tradition: American Schoolbooks of the Nineteenth Century*. Lincoln: University of Nebraska Press, 1964.

Englehardt, N. L. *Toward New Frontiers of Our Global World*. New York: Noble and Noble, 1943.

Fairbanks, Harold W. "Physiography: An Elementary Science Course in the High School." *Journal of Geography* 7 (1909): 217–26.

———. "Physical Geography versus General Science." *School Science and Mathematics* 10 (December 1910): 761–62.

———. *Real Geography and Its Place in the School*. San Francisco: Harr Wagner Publishing Company, 1927.

Fellmann, Jerome D. "The Rise and Fall of High-School Economic Geography." *Geographical Review* 76 (1986): 424–37.

Fenneman, Nevin M. "The Circumference of Geography." *Annals of the Association of American Geographers* 9 (1919): 3–11.

Finch, Vernon C. "Geographical Science and Social Philosophy." *Annals of the Association of American Geographers* 29 (1939): 1–28.

Fischer, Eric. "A German Geographer Reviews German Geography." *Geographical Review* 38 (1948): 307–10.

Fisher, Irving, and O. M. Miller. *World Maps and Globes*. New York: Essential Books, 1944.

Fitchet, Duncan M. "100 Years and Rand McNally." *Surveying and Mapping* 16 (1956): 126–32.

FitzGerald, Frances. *America Revised: History Schoolbooks in the Twentieth Century*. Boston: Little, Brown and Company, 1979.

Flack, James Kirkpatrick. *Desideratum in Washington: The Intellectual Community in the Capital City, 1870–1900*. Cambridge, Mass.: Schenckman, 1975.

Foscue, Edwin J. "The Place of Geography in the Senior High School with Special Reference to Texas." *Journal of Geography* 35 (1936): 117–22.

Freeman, T. W. *A Hundred Years of Geography*. Chicago: Aldine Publishing Company, 1961.

———. *A History of Modern British Geography*. London: Longman, 1980.

"From Scratch." *New Yorker* 25 (29 October 1949): 19–20.

Frye, Alex Everett. *Primary Geography*. Boston: Ginn & Company, 1894.

———. *A Complete Geography*. Boston: Ginn & Company, 1895, 1902.

Gelfand, Lawrence. *The Inquiry: American Preparations for Peace, 1917–1919*. New Haven: Yale University Press, 1963.

"Geographers in Pressing Demand for War Work." *Journal of Geography* 17 (1918): 33–34.

"Geographers in the National Defense Program." *The Professional Geographer* 7 (1948).

"Geography in Secondary Schools." *Journal of Geography* 6 (1908): 220.

Glacken, Clarence. *Traces on the Rhodian Shore: Nature and Culture in Western Thought from the Ancient Times to the End of the Eighteenth Century*. Berkeley: University of California Press, 1967.

Glick, Thomas F. "Before the Revolution: Edward Ullman and the Crisis of Geography at Harvard, 1949–1950." Paper presented at the AAG Annual Conference, San Antonio, 26 April 1982.

———. "History and Philosophy of Geography." *Progress in Human Geography* 8 (1984): 275–83; 9 (1985): 424–31; 10 (1986): 267–77; 11 (1987): 405–15; 12 (1988): 441–50; 14 (1990): 120–28.

———. "Before the Revolution: Edward Ullman and the Crisis of Geography at Harvard, 1949–1950." In *Geography in New England,* edited by John E. Harmon and Timothy J. Rickard. New England: St. Lawrence Valley Geographical Society, 1988.

Godlewska, Anne, and Neil Smith, eds. *Geography and Empire*. Oxford: Basil Blackwell, 1994.

Goetzmann, William H. *Army Exploration in the American West, 1803–1863*. New Haven: Yale University Press, 1959.

Goode, J. Paul. "What the War Should Do for Our Methods in Geography." *Journal of Geography* 18 (1919): 179–84.

———. "The Map as a Record of Progress in Geography." *Annals of the Association of American Geographers* 27 (1927): 1–14.

Gottmann, Jean. "Geography and International Relations." *World Politics* 3, no. 2 (January 1951): 153–73.

Gould, Peter. "Geography 1957–77: The Augean Period." *Annals of the Association of American Geographers* 69 (1979): 139–50.

———. "Commentary on *Reflections on Richard Hartshorne's 'The Nature of Geography.'*" *Annals of the Association of American Geographers* 81 (1991): 328–34.

Greenhood, David. *Down to Earth*. New York: Holiday House, 1944.

Gregory, W. M. "Secondary School Geography in the Middle West." *Journal of Geography* 8 (1910): 110–16.

Griffin, Paul F. "The Contribution of Richard Elwood Dodge to Educational Geography." Ph.D. diss., Columbia University, 1952.

Griswold, Guy Brown Jr. "The Promotion of Travel by the Gasoline Industry." M.B.A. thesis, University of Pennsylvania, 1938.

Grosvenor, Gilbert H. "Map Services of the National Geographic Society." In *The Round Earth on Flat Paper*. Washington, D.C.: National Geographic Society, 1947.

———. *The National Geographic Society and Its Magazine*. Washington, D.C.: National Geographic Society, 1957.

Guelke, Leonard. "Intellectual Coherence and the Foundations of Geography." *Professional Geographer* 41 (1989): 123–30.

Guyot, Arnold. *Elementary Geography*. New York: Scribner, Armstrong, and Company, 1868, 1875.

Hall, G. Stanley. "The Ideal School." *Proceedings of the NEA* (1901): 480.

———. *Educational Problems*. New York: Appleton & Co., 1911.

"Hammond Has a Completely New Atlas." *Publishers' Weekly* 156, no. 2 (3 December 1949): 2276.

Hankins, Grace Croyle. *Our Global World: A Brief Geography for the Air Age*. New York: Gregg, 1944.

Harley, J. B. "Maps, Knowledge, and Power." In *The Iconography of Landscape: Essays on the Symbolic Representation, Design and Use of Past Environments*, edited by Dennis Cosgrove and Stephen Daniels, 277–312. Cambridge: Cambridge University Press, 1988.

———. "Deconstructing the Map." *Cartographica* 26, no. 2 (1989): 1–20.

———. "Introduction." In *From Sea Charts to Satellite Images: Interpreting North American History Through Maps*, edited by David Buisseret. Chicago: University of Chicago Press, 1990.

Harley, J. B., and David Woodward. "Why Cartography Needs Its History." *American Cartographer* 16 (1989): 5–16.

Harris, Chauncy D. "The Department of Geography of the University of Chicago in the 1930s and 1940s." *Annals of the Association of American Geographers* 69 (1979): 21–32.

Harris, Neil. "All the World a Melting Pot? Japan at American Fairs, 1876–1904." In *Mutual Images: Essays in American-Japanese Relations*, edited by Akira Iriye, 24–54. Cambridge: Harvard University Press, 1975.

Harris, William T. "The Place of Geography in the Elementary Schools." *Forum* 32 (1901–2): 539–50.

Harrison, Richard Edes. "Making Maps Tell the Truth." *Travel* 80 (1942): 10–13.

———. "The War of the Maps." *Saturday Review of Literature*, 7 August 1943, 24–27.

———. "The Face of One World: Five Perspectives for an Understanding of the Air Age." *Saturday Review of Literature*, 1 July 1944, 5–6.

———. *Look at the World: The FORTUNE Atlas for World Strategy*. New York: Knopf, 1944.

Harrison, Richard Edes, and Robert Strausz-Hupe. "Maps, Strategy, and World Politics." In *Foundations of National Power*, edited by Harold and Margaret Sprout, 90–93. Princeton: Princeton University Press, 1945.

Harrison, Richard Edes, and Hans W. Weigert. "World View and Strategy." In *Compass of the World: A Symposium on Political Geography*, edited by Hans W. Weigert and Vihjalmur Stefansson. New York: Macmillan, 1944.

Hartshorne, Richard. *The Nature of Geography: A Critical Survey of Current Thought in Light of the Past*. Lancaster, Pa.: Association of American Geographers, 1939.

———. "On the Mores of Methodological Discussion in American Geography." *Annals of the Association of American Geography* 38 (1948): 113–25.

———. "'Exceptionalism in Geography' Reexamined." *Annals of the Association of American Geography* 45 (1955): 205–44.

Haskell, Thomas. *The Emergence of Professional Social Science: The American Social Science Association and the Nineteenth-Century Crisis of Authority*. Urbana: University of Illinois Press, 1977.

Hauptman, Laurence M. "Westward the Course of Empire: Geography Schoolbooks and Manifest Destiny, 1783–1893." *Historian* 40 (1978): 423–40.

Henrikson, Alan K.. "The Map as an 'Idea': The Role of Cartographic Imagery during the Second World War." *American Cartographer* 2 (1975): 19–53.

———. "All the World's a Map." *Wilson Quarterly* 3, no. 2 (spring 1979): 165–77.

———. "America's Changing Place in the World: From 'Periphery' to 'Centre'?" In *Centre and Periphery: Spatial Variation in Politics*, edited by Jean Gottmann. Beverly Hills, Calif.: Sage Publications, 1980.

———. "The Geographical 'Mental Maps' of American Foreign Policy Makers." *International Political Science Review* 1, no. 4 (1980): 495–530.

———. "Mental Maps." In *Explaining the History of American Foreign Relations*, edited by Michael J. Hogan and Thomas G. Paterson, 177–92. Cambridge: Cambridge University Press, 1991.

Herb, Guntram Henrik. *Under the Map of Germany: Nationalism and Propaganda, 1918–1945*. London and New York: Routledge, 1997.

Herbst, Jurgen. "Social Darwinism and the History of American Geography." *Proceedings of the American Philosophical Society* 105 (1961): 538–44.

Heske, Henning. "German Geographical Research in the Nazi Period: A Content Analysis of the Major Geography Journals, 1925–1945." *Political Geography Quarterly* 5 (1986): 267–81.

———. "Karl Haushofer: His Role in German Geopolitics and in Nazi Politics." *Political Geography Quarterly* 6 (1987): 135–44.

Hewes, Leslie. "Dissertations in Geography Accepted by Universities in the United States and Canada for the Ph.D., June, 1935, to June, 1946, and Those Currently in Progress." *Annals of the Association of American Geographers* 36 (1946): 215–47.

"High School Geography." *Journal of Geography* 6 (1907): 102–3.

Hill, A. David, and Lisa A. LaPrairie. "Geography in American Education." In *Geography in America*, edited by Gary L. Gaile and Cort J. Willmott, 1–26. Columbus, Ohio: Merrill Publishing Company, 1989.

Hiltman, John W. "As to the Imported Maps: The Harm Done to American Teaching by the Proposed Increase in Tariff." *Publishers' Weekly* 116 (1929): 1940–41.

Hoffman, Elizabeth. "Is It Essential for the American Student to Be Geographically Informed?" *Journal of Geography* 53 (1954): 149–53.

Hofstadter, Richard. *Social Darwinism in American Thought*. Revised ed. New York: George Braziller, 1959.

Holbrook, Florence. *Rand McNally Elementary Geography*. Chicago: Rand McNally, 1901.

Hollinger, David A. *Postethnic America*. New York: Basic Books, 1995.

Hooson, David, ed. *Geography and National Identity*. London: Basil Blackwell, 1994.

Hudson, Brian. "The New Geography and the New Imperialism: 1870–1918." *Antipode* 9, no. 2 (1977): 12–19.

Huntington, Ellsworth. *The Pulse of Asia*. Boston: Houghton Mifflin & Company, 1907.

———. *Civilization and Climate*. New Haven: Yale University Press, 1915.

———. *World-Power and Evolution*. New Haven: Yale University Press, 1919.

———. *The Character of Races: As Influenced by Physical Environment*. New York: Charles Scribner, 1924.

———. *Tomorrow's Children*. Written with the Directors of the American Eugenics Society. New York: J. Wiley and Sons, 1935.

———. *Season of Birth: Its Relation to Human Abilities*. New York: J. Wiley, 1938.

Huntington, Ellsworth, and Summer Cushing. *Principles of Human Geography*. New York: John Wiley and Sons, 1920.

Huntington, Ellsworth, and Martha Ragsdale. *After Three Centuries: A Typical New England Family*. Baltimore: Williams and Wilkins, 1935.

Huntington, Ellsworth, and Leon F. Whitney. *The Builders of America*. New York: William Morrow, 1927.

Hutchinson, Lincoln. "A Plea for a Broader Conception of Economic Geography." *Journal of Geography* 6 (1907): 122–28.

"Is Geography Receiving Sufficient Attention in the Elementary Schools?" *Journal of Geography* 3 (1904): 189–90.

James, Preston E., and Geoffrey J. Martin. *The Association of American Geographers: The First 75 Years, 1904–1979*. Washington, D.C.: AAG, 1978.

Johnson, Alfred Sidney. "How Maps and Atlases Are Made." *Publishers' Weekly* 101 (1922): 1033–34, 1102–4, 1166–69, 1226–28.

Johnson, Douglas. "The Geographic Prospect." *Annals of the Association of American Geographers* 19 (1929): 205–13.

Johnston, R. J. *Philosophy and Human Geography*. London: Edward Arnold, 1983.

Junge, Ruby M. "Geography in the High Schools of Michigan." *Journal of Geography* 50 (1951): 329–34.

Kearns, Gerry. "Closed Space and Political Practice: Frederick Jackson Turner and Halford Mackinder." *Environment and Planning D* 1 (1984): 23–34.

Kelley, Loulie C. "Physical Geography in Secondary Schools." *Journal of Geography* 6 (1908): 188–91.

Kern, Stephen. *The Culture of Time and Space, 1880–1918*. Cambridge: Harvard University Press, 1983.

Kirby, Andrew. "The Great Desert of the American Mind: Concepts of Space and Time and Their Historiographic Implications." In *The Estate of Social Knowledge*, edited by JoAnne Brown and David van Keuren. Baltimore: Johns Hopkins University Press, 1991.

Klimm, Lester E. "The Nature of Geography: A Commentary on the Second Printing." *Geographical Review* 32 (1947): 486.

Knowlton, Daniel C. "The Relation of Geography to the Social Studies in the Curriculum." *Journal of Geography* 20 (1921): 225–34.

Koelsch, William A. "Terrae Incognitae and Arcana Siwash: Toward a Richer History of Academic Geography." In *Geographies of the Mind*, edited by David Lowenthal and Martyn J. Bowden, 63–87. New York: Oxford University Press, 1976.

———. "Wallace Atwood's 'Great Geographical Institute.'" *Annals of the Association of American Geographers* 70 (1980): 567–82.

———. "'Better than Thou': Rating of Geography Departments, 1924–1980." *Journal of Geography* 80 (1981): 164–69.

Kristof, L. K. D. "The Origins and Evolutions of Geopolitics." *Journal of Conflict Resolution* 4 (1960): 15–51.

LaFeber, Walter. *The New Empire: An Interpretation of American Expansion, 1860–1898*. Ithaca: Cornell University Press, 1963. Reprinted 1998.

Lawler, Thomas Bonaventure. *Seventy Years of Textbook Publishing: A History of Ginn and Company, 1867–1937*. Boston: Ginn and Company, 1938.

Lawrence, Chester, ed. *New World Horizons: Geography for the Air Age*. New York: Duell, Sloan and Pearce, 1942.

Lawrence, Paul F. "The Status of Geography in Secondary Schools of New Jersey." M.A. thesis, Stanford University, 1946.

Leigh, Ruth. "Selling Globes and Atlases." Reprinted from *Publishers' Weekly*. Chicago: Rand McNally & Company, 1929.

Leighly, John. "Some Comments on Contemporary Geographic Method." *Annals of the Association of American Geographers* 27 (1937): 125–41.

———. "What Has Happened to Physical Geography?" *Annals of the Association of American Geographers* 45 (1955): 309–18.

———. "Drifting into Geography in the 1920s." *Annals of the Association of American Geographers* 69 (1979): 4–9.

Lewis, Martin, and Kären Wigen. *The Myth of Continents: A Critique of Metageography*. Berkeley: University of California Press, 1997.

Lewthwaite, Gordon. "Environmentalism and Determinism: A Search for Clarification." *Annals of the Association of American Geographers* 56 (1966): 1–23.

Littlefield, Henry Miller. "Textbooks, Determinism, and Turner's Westward Movement in Secondary School History and Geography Textbooks, 1830–1960." Ph.D. diss., Columbia University, 1967.

Livingstone, David. "The History of Science and the History of Geography: Interactions and Implications." *History of Science* 22 (1984): 271–302.

———. "Evolution, Science and Society: Historical Reflections on the Geographical Experiment." *Geoforum* 16 (1985): 119–30.

———. "Geography, Tradition and the Scientific Revolution: An Interpretative Essay." *Transactions of the Institute of British Geographers*, n.s., 15 (1990): 359–73.

———. "The Moral Discourse of Climate: Historical Considerations on Race, Place and Virtue." *Journal of Historical Geography* 17 (1991): 413–34.

———. *The Geographical Tradition: Episodes in the History of a Contested Enterprise.* London: Basil Blackwell, 1992.

———. "A Geologist by Profession, a Geographer by Inclination: Nathanial Southgate Shaler and Geography at Harvard." In *Science at Harvard*, edited by C. Elliott. Cambridge: Harvard University Press, 1992.

Livingstone, David, and J. A. Campbell. "Neo-Lamarckism and the Development of Geography in the United States and Great Britain." *Transactions of the Institute of British Geographers*, n.s., 8 (1983): 267–94.

Lochhead, Elspeth Nora. "The Emergence of Academic Geography in Britain in its Historical Context." Ph.D. diss., University of California, Berkeley, 1980.

Lowenthal, David. "Geography, Experience, Imagination: Towards a Geographical Epistemology." *Annals of the Association of American Geographers* 51 (1961): 241–60.

Lutz, Catherine A., and Jane L. Collins. *Reading National Geographic.* Chicago: University of Chicago Press, 1993.

Luyten, W. J. "Those Misleading New Maps." *Harper's Magazine* 187 (October 1943): 447–49.

Mackinder, Halford J. "On the Scope and Methods of Geography." *Proceedings of the Royal Geographical Society* 9 (1887): 141–60.

Mackinder, Halford J. "The Geographical Pivot of History." *Geographical Journal* 23 (1904): 421–37.

Mackinder, Halford J. "The Round World and the Winning of the Peace." *Foreign Affairs* 21 (July 1943): 595–605.

Mahan, Alfred Thayer. *The Influence of Sea Power upon History, 1660–1783.* Boston: Little, Brown and Company, 1890.

Malin, James C. "Space and History: Reflections on the Closed-Space Doctrines of Turner and Mackinder and the Challenge of Those Ideas by the Air Age." *Agricultural History* 18 (April and July 1944): 65–74 and 107–26.

Manchester, Harland. "The World of Rand McNally." *Saturday Review of Literature*, 21 January 1950, 35–37.

"Maps in Great Demand." *Publishers' Weekly* 141 (3 January 1942): 31–32.

"Maps." *Consumer Reports* 8, no. 11 (30 November 1943): 291–95.

"Maps: Global War Teaches Global Cartography." *Life*, 3 August 1942, 57.

Marsh, George Perkins. *Man and Nature, or Physical Geography as Modified by Human Action*. New York: Charles Scribner, 1864.

Marshall, Leon C. "The Proposal of the Commission of the Association of Collegiate Schools of Business." In *The Twenty Second Yearbook of the National Society for the Study of Education*, Part 2: *The Social Studies*, edited by Guy Montrose Whipple. Bloomington, Ill.: Public School Publishing Company, 1923.

Martin, A. F. "The Necessity for Determinism." *Transactions and Papers of the Institute of British Geographers* no. 17 (1951).

Martin, Geoffrey. *Mark Jefferson: Geographer*. Ypsilanti: Eastern Michigan University Press, 1968.

———. *Ellsworth Huntington: His Life and Thought*. Hamden, Conn.: Archon Books, 1973.

———. "Paradigm Change: A Study in the History of Geography in the United States, 1892–1925." *National Geographic Research* 1 (1985).

———. "The Emergence and Development of Geographic Thought in New England." *Economic Geography* 1998 (Special Issue): 1–13.

Martin, Geoffrey J., and Preston E. James. *All Possible Worlds: A History of Geographical Ideas*. New York: John Wiley, 1993 [1972].

Maury, Matthew Fontaine. *Physical Geography*. New York: Harper, 1868.

Mayer, Max, "Maps and Their Making." *Publishers' Weekly* 118 (1930).

Mayo, William C. "Development of Secondary School Geography as an Independent Subject in the U.S. and Canada." Ph.D. diss., University of Michigan, 1964.

McBryde, F. Webster. "Origin of the American Society for Professional Geographers: Take-Over and Cover-Up by Association of American Geographers Number 1." In *The American Society for Professional Geographers: Papers Presented on the Occasion of the Fiftieth Anniversary of its Founding*. Occasional Publications of the AAG, no. 3 (April 1993).

McCannon, John. "To Storm the Arctic: Soviet Polar Exploration and Public Visions of Nature in the USSR, 1932–1939." *Ecumene* 2, no. 1 (1995): 15–31.

McConnell, William R. *The United States in the Modern World*. Chicago: Rand McNally, 1932.

McMurry, Frank M., and A. E. Parkins. *World Geography: Book II, The Old World*. New York: Macmillan and Company, 1925.

———. *World Geography: Book I, The New World*. New York: Macmillan and Company, 1927.

McNally III, Andrew. *The World of Rand McNally*. New York: Newcomen Society of America, 1956.

Means, Margaret. "International Problems: A Study in Political Geography." *Journal of Geography* 34 (1935): 187–92.

Mensoian, Michael G. Jr. "Development of Geography Textbooks from 1800–1950." M.A. thesis, Boston University, 1951.

Meyer, Jacob G., and O. Stuart Hamer. *The Old World and Its Gifts*. Chicago: Follet, 1942.

Mikesell, Marvin W. "The Borderlands of Geography as a Social Science." In *Interdisciplinary Relationships in the Social Sciences*, edited by Muzafer Sherif and Carolyn W. Sherif. Chicago: Aldine Publishers, 1969.

Miller, E. Willard. "Geography in the Army Specialized Training Programme." *The Professional Geographer* 3, no. 3/4 (May–June 1945).

Miller, George. "National Council of Geographic Education." *Journal of Geography* 19 (1920): 69–76.

Miller, Rex C. "High School Geography in Nebraska." *Journal of Geography* 47 (1948): 8–17.

Mitchell, Samuel A. *Mitchell's Geographical Reader, A System of Modern Geography*. Philadelphia: E. H. Butler & Company, 1840.

————. *A System of Modern Geography, Comprising a Description of the Present State of the World, and Its Grand Divisions*. Philadelphia: E. H. Butler & Company, 1873.

————. *First Lessons in Geography: For Young Children*. Philadelphia: E. H. Butler, 1874.

Monmonier, Mark. "Maps in *The New York Times*, 1860–1980: A Study in the History of Journalistic Cartography." *Proceedings of the Pennsylvania Academy of Science* 58 (1984): 79–83.

————. "Maps in *The Times* (of London) and *The New York Times*, 1870–1980: A Cross-National Study in Journalistic Cartography." *Proceedings of the Pennsylvania Academy of Science* 59 (1985): 61–66.

————. *Maps with the News: The Development of American Journalistic Cartography*. Chicago: University of Chicago Press, 1989.

Morris, Rita Mary. "Examination of Some Factors Related to the Rise and Decline of Geography as a Field of Study at Harvard, 1638–1948." Ph.D. diss., Harvard University, 1962.

Morse, Jedediah. *American Universal Geography, or, A View of the Present State of All the Empires, Kingdoms, States, and Republics of the Known World, and of the United States of America in Particular*. Originally published by Shepard Kollock, for the author, 1789.

————. *Geography Made Easy*. Boston: J. T. Buckingham, 1806.

Mott, Frank Luther. *A History of American Magazines*. Cambridge: Harvard University Press, 1930–1968.

Mulhern, James. *A History of Secondary Education in Pennsylvania*. Lancaster, Pa.: Science Press, 1933.

Murra, Wilker Finn, et al. *Bibliography of Textbooks in the Social Studies for Elementary and Secondary Schools*. National Council for the Social Studies Bulletin no. 12. Cambridge, Mass.: Samuel Marcus Press, 1939.

National Education Association. *Report of the Committee on Secondary School Studies*. Washington, D.C.: Government Printing Office, 1893.

————. "Report of the Committee on Physical Geography." In *Addresses and Proceed-*

ing of the NEA, vol. 38. Organized under the Report of the Committee on College-Entrance Requirements. Chicago: University of Chicago, 1899.

————. *The Social Studies in Secondary Education*. A Report of the Committee on Social Studies on the Reorganization of Secondary Education of the NEA. *Bulletin*, no. 28. Washington, D.C.: Bureau of Education, 1916.

National Geographic Magazine, 1888–1929.

National Society for the Study of Education. *The Teaching of Geography*. The Thirty-Second Yearbook of the National Society for the Study of Education. Bloomington, Ill.: NSSE, 1933.

"New Atlases in Demand." *Publishers' Weekly* 118 (1930): 2198.

Nida, Stella Humphrey. *Panama and Its Bridge of Water*. Chicago: Rand McNally, 1915.

Nietz, John. *Old Textbooks: Spelling, Grammar, Reading, Arithmetic, Geography, American History, Civil Government, Physiology, Penmanship, Art, Music, as Taught in the Common Schools from Colonial Days to 1900*. Pittsburgh: University of Pittsburgh Press, 1961.

————. *Evolution of American Secondary School Textbooks*. Rutland, Vt.: C. E. Tuttle, 1966.

Odell, Clarence Burt, and Leslie Wood White. "The Status of Geography in the High Schools of Missouri." *Journal of Geography* 41 (1942): 41–51.

Ohmann, Richard. *Selling Culture: Magazines, Markets, and Class at the Turn of the Century*. London and New York: Verso, 1996.

Ó Tuathail, Gearóid. "'It's Smart to Be Geopolitical': Narrating German Geopolitics in United States Political Discourse, 1939–1943." Paper presented at the annual meeting of the Association of American Geographers, San Francisco, March 29–April 1, 1994. Reprinted in *Critical Geopolitics*. Minneapolis: University of Minnesota Press, 1996.

Packard, Leonard O. "Geography and Reconstruction in Education." *Journal of Geography* 18 (1919): 24–28.

Packard, Leonard, with Bruce Overton and Charles Sinnott. *Nations at Work: An Industrial and Commercial Geography*. New York: Macmillan, 1933.

Packard, Leonard, with Bruce Overton and Ben Wood. *Our Air-Age World: A Textbook in Global Geography*. New York: Macmillan, 1944.

Packard, Leonard, with Bruce Overton and Ben Wood. *Geography of the World for High Schools*. New York: Macmillan, 1948.

Packard, Leonard, and Charles Sinnott. *Nations as Neighbors, a Textbook in Geography for Junior High Schools and for Classes of Corresponding Grades*. New York: Macmillan, 1925.

Paddock, Miner H. "Physical Geography in Our Public Schools." *Education* 25 (1904): 162–63.

Parkins, A. E. "The Geography of American Geographers." *Journal of Geography* 33 (September 1934): 221–30.

Parsons, J. J. "The Later Sauer Years." *Annals of the Association of American Geographers* 69 (1979): 9–15.

Pauly, Philip J. "The World and All That Is In It: The National Geographic Society, 1888–1918." *American Quarterly* 31 (1979): 517–32.

———. "The Development of High School Biology: New York City, 1900–1925." *Isis* 82 (1991): 662–88.

"Perspective Maps: Harrison Atlas Gives Fresh New Look to Old World." *Life*, 28 February 1944, 56–61.

Peters, Cynthia Huggins. "Rand McNally and Company—Printers, Publishers, Cartographers: A Study in Nineteenth-Century Mass Marketing." M.L.S. diss., University of Chicago, 1981.

"Physical Geography versus General Science." *School Science and Mathematics* 10 (1910): 761–72.

Pickles, John. "Text, Hermeneutics, and Propaganda Maps." In *Writing Worlds: Discourse, Text, and Metaphor in the Representation of Landscape*, edited by Trevor Barnes and James Duncan. London: Routledge, 1992.

"President's Speech Causes Great Demand for Maps." *Publishers' Weekly* 141 (1942): 935–36.

Quam, Louis O. "The Use of Maps in Propaganda." *Journal of Geography* 42 (1943): 21–32.

Raisz, Erwin. "Outline of the History of American Cartography." *Isis* 26 (1937): 373–89.

———. *General Cartography*. New York: McGraw-Hill, 1948.

"Rand McNally & Company—Global Briefing." *New Yorker*, 4 May 1946, 17–18.

"Rand McNally & Company—War and Map Makers." *Newsweek*, 1 January 1940, 26.

"Rand McNally Wins Carey-Thomas Award for World Atlas." *Publishers' Weekly* 157 (4 February 1950): 783–87.

Rand McNally. *Catalog of Maps, Globes, and Atlases for Schools*. Chicago: Rand McNally, 1912.

———. *Catalog of Maps and Globes for Schools*. Chicago: Rand McNally, 1922.

———. *Goode Political and Physical Maps*. Chicago: Rand McNally, c. 1922.

———. *Catalog of Educational Publications*. Chicago: Rand McNally and Company, 1928.

———. "For Increasing Duties on Geographical Maps." Reply brief of Rand McNally to D. Appleton & Company, Committee of Ways and Means, Tariff Readjustment, United States Congress, 1929, pp. 1–4.

———. "Air Education Adds a Third Dimension to Your Curriculum." Chicago: Rand McNally, c. 1942.

Raup, H. P. "The Impact of Air Power on Geographical Education." *Journal of Geography* 46 (1957): 117–25.

Redman, Anabel. *Classified Catalogue of Textbooks in the Social Studies for Elementary and Secondary Schools*. National Council of the Social Studies Publication no. 2. New York: The Bristol Press, 1927.

Redway, Jacques W. *Commercial Geography: A Book for High Schools*. 1903. New York: Charles Scribner's Sons.

Redway, Jacques W., and Russell Hinman. *Natural Advanced Geography*. New York: American Book Co., 1897.

———. *Natural Elementary Geography*. New York: American Book Co., 1897.

Renner, George T. "The Geography Curriculum." *Journal of Geography* 29 (1930): 344–53.

———. *Human Geography for the Air Age*. New York: Macmillan, 1942.

———. "Maps for a New World." *Collier's*, 6 June 1942, 14–28.

———. "Air Age Geography." *Harper's Magazine* 187 (June 1943): 38–41.

———. "Peace by the Map." *Collier's*, 4 June 1944, 44–47.

———. "What the War has Taught Us About Geography." *Journal of Geography* 43 (1944): 8.

Renner, George T. *Global Geography*. New York: Thomas Crowell, Co., 1944.

Rice, Joseph Mayer. *The Public-School System of the United States*. New York: The Century Company, 1893.

Ridgway, Henry W. "A Study of Representative Commercial Geography Textbooks, 1905–1925." Ph.D. diss., University of Pennsylvania, 1930.

Ristow, Walter W. "Air Age Geography: A Critical Appraisal and Bibliography." *Journal of Geography* 43 (1944): 337–43.

———. "Journalistic Cartography." *Surveying and Mapping* 27 (October–December 1957): 369–90.

———. "Lithography and Maps, 1796–1850." In *Five Centuries of Map Printing*, edited by David Woodward. Chicago: University of Chicago Press, 1975.

———. *Maps for an Emerging Nation: Commercial Cartography in Nineteenth-Century America*. Washington, D.C.: Library of Congress, 1977.

———. "The French-Smith Map and Gazetteer of New York State." *Quarterly Journal of the Library of Congress* 36, no. 1 (winter 1979): 68–90.

———. *American Maps and Mapmakers: Commercial Cartography in the Nineteenth Century*. Detroit: Wayne State University Press, 1985.

Robinson, Arthur. "Mapmaking and Map Printing: The Evolution of a Working Relationship." In *Five Centuries of Map Printing*, edited by David Woodward. Chicago: University of Chicago Press, 1975.

Robinson, Edward Van Dyke. *Commercial Geography*. Chicago: Rand McNally, 1922 [1915, 1920].

Roorbach, George B. "The Trend of Modern Geography, A Symposium." *Bulletin of the American Geographical Society* 46 (November 1914): 798–808.

Rosen, Sidney. "A Short History of High School Geography to 1936." *Journal of Geography* 56 (1957): 405–13.

Rosenberg, Charles E. "Catechisms of Health: The Body in the Prebellum Classroom." *Bulletin of the History of Medicine* 69 (1995): 175–97.

Rosenberg, Emily. *Spreading the American Dream: American Economic and Cultural Expansion, 1890–1945*. New York: Hill and Wang, 1982.

Ross, Dorothy. *The Origins of American Social Science*. New York: Cambridge University Press, 1991.

Rostlund, Erhard. "Twentieth-Century Magic." In *Readings in Cultural Geography*, edited by Marvin Mikesell. Chicago: University of Chicago Press, 1962.

Rothenberg, Tamar. "*National Geographic*'s World: The Politics of Popular Geography, 1888–1945." Ph.D. diss., Rutgers University, 1999.

Rowell, P. E. "General Science vs. Physical Geography." *School Science & Mathematics* 11 (1911): 116–21.

Rowley, Virginia M. *J. Russell Smith: Geographer, Educator, and Conservationist*. Philadelphia: University of Pennsylvania Press, 1964.

Rugg, Earl. "How the Current Courses in History, Geography, and Civics Came to Be what They Are." *The Social Studies in the Elementary and Secondary School*. The Twenty-Second Yearbook of the National Society for the Study of Education. Bloomington, Ill.: NSSE, 1923.

Ruiz, Ernesto A. "Geography and Diplomacy: The American Geographical Society and the 'Geopolitical' Background of American Foreign Policy, 1848–1861." Ph.D. diss., Northern Illinois University, 1975.

Russell, Israel C. "An American Geographical Society." *Science*, n.s., 15, no. 370 (31 January 1902): 195.

Russell, J. E. "History and Geography in the Higher Schools of Germany." *School Review* 5 (1897): 257–68, 539–47.

Ryan, James R. *Picturing Empire: Photography and the Visualization of the British Empire*. London: Reaktion, 1997. Reprinted under Chicago: University of Chicago Press, 1997.

Rydell, Robert. *All the World's a Fair: Visions of Empire at American International Expositions, 1876–1916*. Chicago: University of Chicago Press, 1984.

Said, Edward. *Orientalism*. New York: Viking, 1978.

Salisbury, Rollin G. "The Teaching of Geography—A Criticism and a Suggestion." *Journal of Geography* 8 (1909): 49–55.

———. "Physiography in the High School." *Journal of Geography* 9 (1910): 57–63.

Sauer, Carl O. "The Survey Method in Geography and Its Objectives." *Annals of the Association of American Geographers* 14 (1924): 17–33.

———. "The Morphology of Landscape." *University of California Publications in Geography* 2, no. 2 (1925). Reprinted in Sauer, *Land and Life*, edited by John Leighly. Berkeley: University of California Press, 1963.

———. "Forward to Historical Geography." *Annals of the Association of American Geographers* 31 (1941): 1–24.

Saxon, Gerald. "How the West was Taught." Paper presented at the History of Cartography Meeting, Newberry Library, Chicago, 1993.

Schaefer, Fred K. "Exceptionalism in Geography: A Methodological Examination," *Annals of the Association of American Geographers* 43 (1953): 226–46.

Scharfe, Wolfgang. "German Atlas Development during the Nineteenth Century." In *Images of the World: The Atlas through History*, edited by John Wolter. Forthcoming.

Schlesinger, Arthur. *The Cycles of American History*. Boston: Houghton Mifflin, 1986.

Schneirov, Richard. *The Dream of a New Social Order: Popular Magazines in America, 1893–1914*. New York: Columbia University Press, 1994.

Schwartz, Joan. "*The Geography Lesson:* Photographs and the Construction of Imaginative Geographies." *Journal of Historical Geography* 22, no. 1 (1996): 16–45.

Semple, Ellen Churchill. *American History and Its Geographic Conditions*. Boston: Houghton Mifflin, 1903.

————. "Emphasis upon Anthropogeography in Schools." *Journal of Geography* 3 (1904): 366–74.

Sereiko, George. "Chicago and its Book Trade, 1871–1893." Ph.D. diss., Case Western Reserve University, 1973.

Seymour, Charles. *Geography, Justice, and Politics at the Paris Conference of 1919*. New York: American Geographical Society, 1951.

Shatto, Theodore C. "Geography in the Senior High Schools of California." M.A. thesis, Stanford University, 1947.

Shepherd, Edith P. *Geography for Beginners*. Chicago: Rand McNally, 1921.

Silk, Leonard, and Mark Silk. *The American Establishment*. New York: Avon Books, 1981.

Sizer, Theodore. *Secondary Schools at the Turn of the Century*. New Haven: Yale University Press, 1964.

Smith, J. Russell. "Geography and the Higher Citizenship." *Progressive Education* 11 (June 1925): 77–80.

————. *Commerce and Industry*. New York: Henry Holt, 1925.

————. *Foreign Lands and Peoples*. Philadelphia: John C. Winston, 1933.

————. *Human Geography. Book 2: Regions and Trade*. Philadelphia: John C. Winston, 1926 [1921].

————. *Our Industrial World*. Philadelphia: John C. Winston, 1934.

————. "American Geography: 1900–1904." *Professional Geographer* 4 (1952): 4–7.

Smith, Neil. "Isaiah Bowman: Political Geography and Geopolitics." *Political Geography Quarterly* (January 1984): 69–76.

————. "Bowman's *New World* and the Council on Foreign Relations." *Geographical Review* 76 (1986): 438–60.

————. "Academic War over the Field of Geography: The Elimination of Geography at Harvard, 1947–1951." *Annals of the Association of American Geographers* 77 (1987): 155–72.

————. "For a History of Geography: Response to Comments." *Annals of the Association of American Geography* 78 (1988): 159–63.

————. "Geography as Museum: Private History and Conservative Idealism in *The Nature of Geography*." In *Reflections on Richard Hartshorne's* The Nature of Geography, edited by Neil Entrikin. Washington, D.C.: Association of American Geographers, 1989.

————. *The Geographic Pivot of History: Isaiah Bowman and the Geography of the American Century.* Baltimore: Johns Hopkins University Press, forthcoming.

Smith, Neil, and Anne Godlewska, eds. *Geography and Empire.* Oxford: Basil Blackwell, 1994.

Smith, Thomas R., and Lloyd D. Black. "German Geography: War Work and Present Status." *Geographical Review* 36 (1946): 398–408.

Snyder, John P. *Flattening the Earth: Two Thousand Years of Map Projections.* Chicago: University of Chicago Press, 1993.

Somerville, Mary. *Physical Geography.* 2d ed. Philadelphia: Lea & Blanchard, 1850.

Sparhawk, Frances Campbell. *Miss West's Class in Geography.* Boston: Lee & Shepard, 1887.

Spate, O. H. K. "How Determined is Possibilism?" *Geographical Studies* 4 (1957): 3–12.

Speier, Hans. "Magic Geography." *Social Research* 8 (1941): 311.

Spieseke, Alice Winifred. *Bibliography of Textbooks in the Social Studies.* National Council for the Social Studies Bulletin no. 23. Menasha, Wis.: George Banta, 1949.

Sprout, Harold, and Margaret Sprout. *Foundations of National Power.* Princeton: Princeton University Press, 1945.

Spykman, Nicholas J., and Helen R. Nicholl, eds. *The Geography of the Peace.* New York: Harcourt Brace, 1944.

Staples, Z. Carleton, and G. Morrell York. *Economic Geography.* New York: Southwest Publishing, 1940.

Steel, Robert W. "The Beginning and the End." In *British Geography, 1918–1945,* edited by R. W. Steel. Cambridge: Cambridge Univerity Press, 1987.

Steinberg, Theodore. "New England in a Pocketbook: Gazetteers and the Modernization of Landscape." *American Studies* 35, no. 2 (1994): 59–72.

Stocking, George W. Jr. "Lamarckianism in American Social Science: 1890–1915." *Journal of the History of Ideas* 23 (1962): 239–40.

Stoddart, David. "Darwin's Impact on Geography." *Annals of the Association of American Geographers* 56 (1966): 683–98.

Stoddart, David. *On Geography and Its History.* New York: Basil Blackwell, 1986.

Stone, J. C. "Imperialism, Colonialism and Cartography." *Transactions of the Institute of British Geographers,* n.s., 13 (1988): 57–64.

Stone, Kirk H. "Geography's Wartime Service." *Annals of the Association of American Geographers* 69 (1979): 89–96.

Stout, Cyril. "Trends, Methods, Contents, and Beliefs in Geography Textbooks, 1784–1895." Ph.D. diss., George Peabody College for Teachers, 1937.

Stout, John Elbert. *The Development of High-School Curricula in the North Central States from 1860 to 1918.* Chicago: University of Chicago Press, 1921.

Stowers, Dewey M. "Geography in American Schools, 1892–1935." Ed.D. diss., Duke University, 1962.

Strong, Josiah. *Expansion under New World-Conditions.* New York: Garland Publishing, 1971 [1900].

Stull, DeForest, and Roy W. Hatch. *Our World Today: Asia, Latin America, United States*. Boston: Allyn and Bacon, 1948.

"The Sudden Demand for Maps." *Publishers' Weekly* 136, no. 1 (9 September 1939): 881–83.

Sutherland, William J. "The Rational Element as an Organizing Principle in Geography." *Journal of Geography* 4 (1905): 97–105.

———. *The Teaching of Geography*. Chicago: Scott, Foresman and Company, 1909.

Swartz, Jon. *Handbook of Old-Time Radio*. Metuchen, N.J.: Scarecrow, 1993.

Swinton, William. *A Complete Course in Geography: Physical, Industrial, and Political*. New York: Ivison, Blakeman, & Taylor, 1875.

Tarr, Ralph S. "Physical Geography—The Teacher's Outfit." *School Review* 4 (1896): 161–72, 193–201.

Tarr, Ralph S., and Frank McMurry. *A Complete Geography*. New York: Macmillan, 1904.

———. *New Geographies, Second Book*. New York: Macmillan, 1919 [1910] .

Tarr, Ralph S., with Ray Whitbeck, Martha Krug-Genthe, and Mark Jefferson. "Results to be Expected." *Journal of Geography* 4 (1905): 145–63.

Taylor, Griffith, ed. *Geography in the Twentieth Century: A Study of Growth, Fields, Techniques, Aims and Trends*. New York: Philosophical Library, 1951.

Tebbel, John, and Mary Ellen Zuckerman, *The Magazine in America, 1741–1990*. New York: Oxford University Press, 1991.

Teggart, Frederick J. "Geography as an Aid to Statecraft: An Appreciation of Mackinder's 'Democratic Ideals and Reality.'" *Geographical Review* 8 (1919): 227–42.

———. "Human Geography: An Opportunity for the University." *Journal of Geography* 18 (1919): 142–48.

Thomas, Helen Goss. "The New Geography: Education for World Citizenship." *Educational Review* 59 (March 1920): 236–43.

Thomas, Louis B. "Maps as Instruments of Propaganda." *Surveying and Mapping* 9 (1949): 75–81.

Thomas, Olive J. "Development and Presentation and Organization of Elementary-School Geography in the United States." *Journal of Geography* 15 (1917): 213–21.

Thompson, Morris. *Maps for America: Cartographic Products of the U.S. Geological Survey and Others*. Washington, D.C.: U.S. Government Printing Office, 1979.

Thralls, Zoe A., and Edwin H. Reeder. *Geography in the Elementary School*. Chicago: Rand McNally, 1931.

Trotter, Spencer. *The Geography of Commerce: A Textbook*. New York: Macmillan, 1911.

Tuathail, Gearóid Ó. *See* Ó Tuathail, Gearóid

Turk, Richard W. *The Ambiguous Relationship: Theodore Roosevelt and Alfred Thayer Mahan*. New York: Greenwood, 1987.

Turner, Frederick Jackson. "Geographical Interpretations of American History." *Journal of Geography* 4 (1905): 34–37.

———. "The Significance of the Frontier in American History." In *The Frontier in American History*. New York: Henry Holt, 1920.

U.S. Bureau of Education. *Cardinal Principles of Secondary Education*. Bulletin no. 35. Washington, D.C.: U.S. Government Printing Office, 1928 [1918].

U.S. Bureau of Education. "Recent Developments in the Teaching of Geography in Europe." In *Report of the Commissioner of Education for the Year 1892–1893*, vol. 1. Washington, D.C.: U.S. Government Printing Office, 1895.

van Loon, Hendrik Willem. *Van Loon's Geography: The Story of the World We Live In*. New York: Simon and Schuster, 1932.

Vining, James W. *The National Council for Geographic Education: The First Seventy-Five Years and Beyond*. Indiana, Pa.: NCGE, 1990.

von Engeln, O. D. "A Campaign for Geography." *Journal of Geography* 18 (1919): 28–31.

Walworth, Arthur. *America's Moment, 1918: American Diplomacy at the End of World War One*. New York: W. W. Norton, 1977.

The War in Maps: An Atlas of New York Times Maps, text by Francis Brown. New York: Oxford University Press, 1944.

"War Services of Members of the AAG." *Annals of the Association of American Geographers* 9 (1919): 49–70.

Warren, David. *Common School Geography* and *System of Physical Geography*. Philadelphia: Cowperthwait, 1868–1879.

Weigert, Hans W., and Vihjalmur Stefansson. *Compass of the World: A Symposium on Political Geography*. New York: Macmillan, 1944.

West, Edith, with Dorothy Meredith and Edgar B. Wesley. *Contemporary Problems Here and Abroad*. Boston: D. C. Heath and Company, 1947.

"What Is Geography?" *Journal of Geography* 10 (1911): 59–60.

Whitbeck, Ray Hughes, with Loyal Durand Jr. and Joe Russell Whitaker. *The Working World*. New York: American Book Company, 1937.

Whittemore, Katheryne T. "Celebrating 75 years of the *Journal of Geography*, 1897–1972." *Journal of Geography* 71 (1972): 7–18.

Whittlesey, Derwent. "Dissertations in Geography Accepted by Universities in the United States for the Degree of Ph.D. as of May, 1935." *Annals of the Association of American Geographers* 25 (1935): 211–37.

Whittlesey, Derwent. *German Strategy of World Conquest*. New York: Farrar & Rinehart, 1942.

Whittlesey, Derwent. "The Horizon of Geography." *Annals of the Association of American Geographers* 35 (1945): 1–36.

"Who Will Sell Maps?" *Publishers' Weekly* 119 (1931): 2310.

Wilford, John Noble. *The Mapmakers: The Story of the Great Pioneers in Cartography from Antiquity to the Space Age*. New York: Vintage Books, 1982.

Williams, William Appleman. *The Tragedy of American Diplomacy*. New York: Dell Publishing, 1962.

Wilson, Leonard S. "Some Observations on Wartime Geography in England." *Geographical Review* 36 (1946): 597–612.

———. "Library Filing, Classification, and Cataloging of Maps, with Special Reference to Wartime Experience." *Annals of the Association of American Geographers* 38 (1948): 6–37.

———. "Lessons from the Experience of the Map Information Section, OSS." *Geographical Review* 39 (April 1949): 298–310.

Winichakul, Thongchai. *Siam Mapped: A History of the Geo-Body of a Nation.* Honolulu: University of Hawaii Press, 1994.

Wohl, Robert. *A Passion for Wings: Aviation and the Western Imagination, 1900–1918.* New Haven: Yale University Press, 1994.

Wolf, Armin. "What Can the History of Historical Atlases Teach? Some Lessons from a Century of Putzger's *Historioscher Schul-Atlas.*" *Cartographica* 28, no. 2 (1991): 21–37.

Wood, Denis. "Pleasure in the Idea: The Atlas as Narrative Form." *Cartographica* 24, no. 1 (1987): 24–45.

———. *The Power of Maps.* New York: Guilford Press, 1992.

Wood, Denis, and John Fels. "Designs on Signs/Myth and Meaning in Maps." *Cartographica* 23, no. 3 (1986): 54–103.

Woodward, David. *The All-American Map: Wax Engraving and Its Influence on Cartography.* Chicago: University of Chicago Press, 1977.

Woodworth, Harmon H. "Now is the Time to Sell Atlases." *Publishers' Weekly* 119 (1931): 835–38.

Wooldridge, S. W. *The Geographer as Scientist: Essays on the Scope and Nature of Geography.* London: Thomas Nelson and Sons, 1956.

"Words versus Things in Geography Teaching." *Journal of Geography* 6 (1907): 28–29.

Wright, John K. "A Plea for the History of Geography." *Isis* 8 (1926): 477–91.

———. *Geography in the Making: The American Geographical Society, 1851–1951.* New York: American Geographical Society, 1952.

Yeomans, Edward. "Geography." *Atlantic Monthly* 125 (Febrary 1920): 167–72.

———. "Human Geography, A Review." *Geographical Review* 11 (1921): 605.

INDEX

Page numbers in boldface refer to illustrations.

Ullman, Edward, 135
United States Armed Forces, 210–12
United States Army Corps of Engineers.
    See Army Corps of Engineers
United States Army Signal Corps, 53,
    137
United States Bureau of Ethnology, 53,
    60
United States Bureau of Reclamation,
    202
United States Bureau of Statistics, 53,
    57
United States Coast and Geodetic Sur-
    vey, 23
United States Department of Agricul-
    ture, 53
United States Department of State, 53,
    210
United States Department of War, 53,
    61, 65, 137, 212
United States Geological Survey, 10, 23,
    53, 55, 60–61, 69, 71, 186
United States Hydrographic Office, 23,
    53
United States Treasury, 53
university geography. See academic
    geography
University of California, 253n. 5, 254n.
    15
University of Chicago, 75–76, 186, 223,
    254n. 15
University of Cincinnati, 254n. 15
University of Michigan, 254n. 15
University of Minnesota, 254n. 15
University of Missouri, 254n. 15
University of Nebraska, 254n. 15
University of Oklahoma, 254n. 15
University of Pennsylvania, 75–76, 83,
    254n. 15
University of Tennessee, 254n. 15
University of Wisconsin, 254n. 15

Van der Grinten, Alphons, 195
Van der Grinten projection. See map
    projections
Van Dyke, Henry, 116–17
Vassar College, 254n. 15

Venezuelan boundary dispute, 66

war atlases, 6–7, 207
    Spanish-American War, 38–44, **42,
    43,** 176, **177,** 181, 206
    World War I, 180–85, **182–83,** 203,
    206–7, 272n. 11
    World War II, 206–9
War of 1812, 106
Ward, Lester Frank, 77
Ward, Robert DeCourcey, 159, 256nn.
    33, 37
Washington Academy of Sciences, 47
Washington University, 254n. 15
wax engraving. See maps, production of
Webster, Noah, 19
Wesley, Edgar B., 139
West, Edith, 139
Westgate, Lewis, 256n. 33
West Indies, 110
Whitbeck, Ray Hughes, 133, 259n. 31,
    261nn. 7, 8
Wigen, Karen, 3, 34, 99
Williams, John Sharp, 168
Willkie, Wendell, 218, 222
Wilson, Woodrow, 89–90, 166, 169,
    184–86
Wood, Ben, 141
Woodbridge, William Channing, 95
Woodward, David, 26, 39, 230
Worcester, Dean, 56
world atlases
    antebellum, 17–21
    comparative, 27, 29, 192, 231, 272n. 2
    consumption, 6–7, 38, 44, 185–86,
        194, 198, 204, 206–7, 227–28,
        237–38
    cost and availability, 17, 21–22, 27–28
    maps
        features, 31–33, 180–84, 187–95,
            229–37
        regional coverage, 28–31, 178–79,
            280nn. 42, 44
    narrative
        expansionism, 7, 39–44, 178–80
        human-environment relationship,
            35–37, 144–46, 179–80

ν